Base Encounters

Anthropology, Culture and Society

Series Editors:
Professor Vered Amit, Concordia University
Doctor Jamie Cross, University of Edinburgh
and
Professor Christina Garsten, Stockholm University

Recent titles:

Base Encounters

The US Armed Forces in South Korea

Elisabeth Schober

www.plutobooks.com

First published 2016 by Pluto Press
345 Archway Road, London N6 5AA

www.plutobooks.com

Copyright © Elisabeth Schober 2016

The right of Elisabeth Schober to be identified as the author of this work has been
asserted by her in accordance with the Copyright, Designs and Patents Act 1988.

British Library Cataloguing in Publication Data
A catalogue record for this book is available from the British Library

ISBN	978 0 7453 3610 7	Hardback
ISBN	978 0 7453 3605 3	Paperback
ISBN	978 1 7837 1770 5	PDF eBook
ISBN	978 1 7837 1772 9	Kindle eBook
ISBN	978 1 7837 1771 2	EPUB eBook

This book is printed on paper suitable for recycling and made from fully managed
and sustained forest sources. Logging, pulping and manufacturing processes are
expected to conform to the environmental standards of the country of origin.

Typeset by Stanford DTP Services, Northampton, England

Simultaneously printed in the European Union and United States of America

To Elisabeth Katschnig-Fasch,
who left way too soon, but stayed long enough to inspire.

Contents

List of Figures

Series Preface

Anthropology is a discipline based upon in-depth ethnographic works that deal with wider theoretical issues in the context of particular, local conditions—to paraphrase an important volume from the series: *large issues* explored in *small places*. This series has a particular mission: to publish work that moves away from an old-style descriptive ethnography that is strongly area-studies oriented, and offer genuine theoretical arguments that are of interest to a much wider readership, but which are nevertheless located and grounded in solid ethnographic research. If anthropology is to argue itself a place in the contemporary intellectual world, then it must surely be through such research.

We start from the question: 'What can this ethnographic material tell us about the bigger theoretical issues that concern the social sciences?' rather than 'What can these theoretical ideas tell us about the ethnographic context?' Put this way round, such work becomes *about* large issues, *set in* a (relatively) small place, rather than detailed description of a small place for its own sake. As Clifford Geertz once said, 'Anthropologists don't study villages; they study *in* villages.'

By place, we mean not only geographical locale, but also other types of 'place'—within political, economic, religious or other social systems. We therefore publish work based on ethnography within political and religious movements, occupational or class groups, among youth, development agencies, and nationalist movements; but also work that is more thematically based—on kinship, landscape, the state, violence, corruption, the self. The series publishes four kinds of volume: ethnographic monographs; comparative texts; edited collections; and shorter, polemical essays.

We publish work from all traditions of anthropology, and all parts of the world, which combines theoretical debate with empirical evidence to demonstrate anthropology's unique position in contemporary scholarship and the contemporary world.

Professor Vered Amit
Dr Jamie Cross
Professor Christina Garsten

Notes on the Text

Excerpts from the introduction and conclusion have previously been published as 'Vil(l)e Encounters: The US Armed Forces in Korea and Entertainment Districts in and near Seoul', in R. Frank, J.E. Hoare, P. Köllner, and S. Pares (eds) *Korea 2011: Politics, Economy and Society*, Korea Yearbook, Vol. 5 (Leiden: Brill), pp. 207–32. An earlier version of chapter 3 has previously been published as '"The Colonized Bodies of Our Women ...': Imaginative and Material Terrains of US Military Entertainment on the Fringes of South Korea", in G. Frerks, R.S. König, and A. Ypeij (eds) *Gender and Conflict: Embodiments, Discourses and Symbolic Practices* (London: Ashgate, 2014), pp. 133–50. A modified version of chapter 6 has been published as "Itaewon's Suspense: Masculinities, Place-making and the US Armed Forces in a Seoul Entertainment District", *Social Anthropology/Anthropologie Sociale* 22(1): 36–51.

Notes on Transliteration

The Romanization of Korean words and names in this book follows the McCune-Reischauer system except for names whose personal orthography is publicly known, or who have requested idiosyncratic spellings. I follow the Korean naming convention of surname followed by given name in case of Korean persons, and the Western convention of given name first, surname second in reference to Korean Americans.

Acknowledgments

This project has been funded by a Marie Curie Early Stage Training Fellowship (Marie Curie SocAnth), as well as by various grants and awards provided by Central European University. Major revisions to the manuscript have been undertaken during a postdoctoral fellowship at University of Oslo, where I am part of the ERC-Advanced-Grant project 'Overheating: The Three Crises of Globalization'.

I would like to thank Don Kalb for his devoted and generous support throughout the years. His competent assistance and constant encouragement helped me greatly to keep working on this project ever since I began my PhD studies at Central European University (CEU) in 2006. Daniel Monterescu and Sophie Day have also given me tremendously important feedback over the years. Furthermore, I am indebted to a number of people who have spent time and energy making comments on various papers, early chapter drafts and provisional sections of my PhD thesis, such as Don Nonini, Calin Goina, Prem Kumar Rajaram, Frances Pine, Michael Herzfeld, Rogers Brubaker, Lisa Law, Jakob Rigi, David Berliner, James Hoare, and Susan Pares. I have also greatly benefitted from conversations with Kim Yeongmi, Elisa Helms, Matteo Fumagali, Erdem Evren, and Chung Heisu. Thanks also to Dan Rabinowitz and all the participants of the CEU SocAnth write-up seminar, who have greatly helped me in the last stages of writing. I have also benefitted from seminars, workshops, and the larger network provided by the Marie Curie SocAnth Doctoral Training School that I was part of—a big thanks to Michael Stewart and all the other faculty, staff, and students of the anthropology departments in Budapest (CEU), Cluj (Babeş-Bolyai University), Halle (Max Planck Institute for Social Anthropology), London (Goldsmiths), and Sibiu (Astra Film Studio) who were involved in this project. I also want to thank my CEU comrades Olena Fedyuk, Neda Deneva, Anca Simionca, Mariya Ivancheva, Luisa Steur, Alexandra Szőke, Ian Cook, Gábor Halmai, Trever Hagen, and Zoltán Dujisin for their friendship, their encouragement, and support over the last decade. I am deeply indebted also to Kim Bogook from Eötvös Loránd (ELTE) University for the endless patience and support with which he encouraged my first feeble attempts at learning the Korean language (which I could then expand upon with the help of countless teachers at Yonsei University and Sookmyoung Women's University's Korean Language Programs, to whom I am also very grateful).

In Berlin, my gratitude goes to my dear friends and colleagues from Korea-Verband e.V., who have supported me in numerous ways. In particular, thank you, Han Nataly Jung-Hwa, for your friendship, support and mentorship. Tsukasa Yajima, Youngsook Rippel Choi, and Yoo Jae-hyun, thank you for your encouragement and friendship. I would also like to thank Pfarrer Hartmut Albruschat and the members of Korea-Arbeitsgruppe at the Berliner Missionswerk, who took the time to give me feedback on a presentation of my project.

Since my move to University of Oslo in January 2013, numerous new debts have been incurred. In particular, I would like to extend my gratitude to Thomas Hylland Eriksen for his incredible support and intellectual input. I have benefitted greatly from conversations with and comments from Chris Hann, Henrik Sindig-Larsen, Mikkel Vindegg, Penny Harvey, Douglas R. Holmes, Georg Frerks, Annelou Ypeij, and Reinhilde Sotiria König, Christian Krohn-Hansen, Lena Gross, Robert Pijpers, Astrid Stensrud, Wim van Daele, Alanna Cant, Anna Tsing, Cathrine Thorleifsson, George Baca, and Lee Ko Woon. I have also found much encouragement in being able to present this book project to the departmental members at the Department of Social Anthropology lunch seminar—thank you, Keir Martin, for organizing this. Thanks also to the participants of the conference on "The Loose Ends of Fieldwork: Emotional Care of the Self in the Ethnography of Violence" at the University of Copenhagen, and the speakers and discussants at the workshop on "Polarization in Divided Societies: Korea in a Global Context" at CEU, Hungary. Thanks also to Vladimir Tikhonov for his support. And in this context, I would also like to sincerely thank David Castle at Pluto Press for his tireless support of this book project of mine. My gratitude also goes to the anonymous reviewers who read first draft versions and to Jamie Cross for his generous and immensely helpful review of the manuscript.

And of course I want to thank the countless people who have helped me during my time in South Korea, many of whom, for reasons of anonymity, I unfortunately cannot name here. I am deeply indebted to my friend Yu CheongHee, to Kim Elli, and Song Ŭn-ae, who have helped me out so many times. My deepest gratitude to the staff workers at Turebang, and in particular to Yu Young-lim, Yu Pok-nim, and Park Sumi for the kindness, patience, and generosity with which they have welcomed me into their offices and introduced me to the world of kijich'on. Thanks a lot also to the people of Peace Network, Seoulidarity, and World Without War, and to the staff of the National Campaign for Eradication of Crimes by US Troops in Korea and of Haet-sal. At Sarangbang, a drop-in shelter run by the organization Magdalena House, I have also found open doors—thanks a lot to Kim Chu Hŭi in particular. Lina Hoshino from Genuine Security, I have greatly enjoyed our collaboration. Sincere thanks also to my friends Karo, Hungying, El Jefe, Rob, Niko, "Crazy Flower", Jayden, the Hongdae park kids, the It'aewŏn crowd, and the men and women I met in kijich'on.

Lastly, I am deeply indebted to my parents, Ingeborg and Franz Josef Schober, for their infinite support throughout the years. And to my husband, Yi Wonho: this book would not have been possible without you. Thank you for being there all along.

1

Introduction

Violent Imaginaries and Base Encounters in Seoul

"A Certain Neighborhood ..."

In mid January 2007, Private Geronimo Ramirez, a then 23-year-old United States (US) soldier deployed in South Korea, was arrested for the repeated rape of a Korean woman in the Seoul entertainment district of Hongdae. Together with another soldier friend of his, that weekend Ramirez had made the one-and-a-half hour ride from his US military base located in Tongduch'ŏn all the way to central Seoul. The team tried unsuccessfully to check into the Dragon Hill Lodge, a military hotel located within the premises of the Yongsan US Army garrison in Seoul that was booked out that evening, and then decided to go to a motel in Hongdae instead. After a night spent drinking and partying, Ramirez's buddy went back to the motel alone, while Ramirez continued to walk through the streets of the neighborhood, pouring down more beers bought from convenience stores nearby. In a deserted area, he encountered a 67-year-old Korean female in the early morning hours, who was on her way home from a cleaning job. Ramirez would beat and rape the woman repeatedly, on the street, in an alley and inside a building, until he was taken in by Korean police forces that had been alerted by the woman's screams. Ramirez, in his public letter of apology, stated that he had no memory of the sexual assault; and he asked the victim not to "think bad of americans [sic] for everyone makes mistakes and this was mine." He added that "I was suppose[d] to go home soon & get married[,] but now i can't[,] i will stay here & pay for my mistakes" (Slavin and Hwang 2007).

When I arrived in Seoul in the fall of the same year,[1] this brutal incident was still much discussed among locals and foreigners alike. Besides fulfilling certain expectations that many proponents of the nationalist left held about GIs,[2] namely that all US military personnel were potential

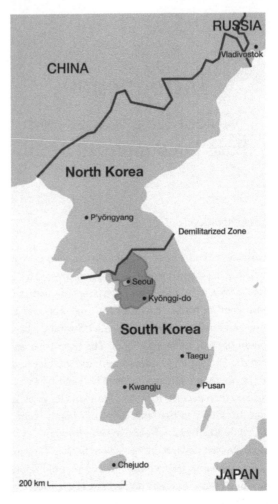

Figure 1.1   Map of the Korean peninsula

perpetrators, the event had also brought to light a recent development that posed a challenge to both US Forces Korea (USFK) and local authorities: many of the nearly 30,000 US soldiers[3] stationed in South Korea no longer seemed to stay in the remote red-light districts close to their base facilities that they had informally been assigned to. These so-called camptowns (*kijich'on* in Korean, also known as "villes" among the soldiers) are entertainment areas catering primarily to US military personnel. The GI bars and clubs in the area are typically run by Korean entrepreneurs who employ a number of female "entertainers" to look after the needs of the US servicemen. They are tightly regulated spaces; the US Military

Police send their own staff to patrol the area and go after US soldiers who are found to be in violation of US or South Korean law. However, now that plenty of servicemen increasingly seemed to party in entertainment districts in central Seoul as well—in downtown neighborhoods often far removed from their bases—the challenge of keeping these young men[4] in line increased disproportionately in difficulty. Many Korean citizens, I was to learn, including those locals left behind in the economically struggling and socially stigmatized areas nearby US bases, would like to contain GIs in the camptowns they emerged from.

I got to know Jay,[5] a 22-year-old US Army member also stationed in Tongduch'ŏn, in late 2007. He had been in Korea for a little under a year, and was about to be relocated to the Middle East over the coming few months. Walking into a popular bar in the downtown district of Chongno with Jay, his Korean girlfriend, and a Korean friend of hers, I became aware of the many stares that the young serviceman, tall, muscular and with short-cropped hair, attracted in this venue. While his friends quietly talked in Korean next to us—politely but decidedly ignoring Jay who would occasionally ask, "What the fuck is it that you are saying?", Jay was entertaining himself by returning some of the stares he received from the neighboring tables until the young Korean people seated there shifted their eyes away. After a while, he started to noisily grind the beer bottle that he had just emptied at the edge of the table we were sitting at, causing additional concerned looks in our direction. He only visibly relaxed when our food arrived; we had ordered grilled chicken, as Jay had ruled out any meal containing *kimchi*,[6] asking me earlier on, "You really eat that shit?"

After some initial remarks by Jay that he would most certainly not be a good conversational partner for me—"I'm not a good guy to talk to, in case you haven't noticed yet. I don't know how to deal with students. I only know how to deal with soldiers, got that?"—Jay began to talk about his life in Tongduch'ŏn where he was stationed. The US military, he argued, invested a lot every year in "good publicity projects," such as sending soldiers out to help with teaching English at Korean schools for a day. "The idea behind this is, of course," Jay added, "that there is already plenty of bad press about us out there." The "ville" of Tongduch'ŏn, he said, was the area that most of his co-workers spent their free time in, going to the bars, clubs, and restaurants catering to their needs.

Asked what his friends did when they had a bit of time to kill, he replied: "Go to whores. Sorry, but that's just how it is. Nothing else to do up there anyways." Filipina "entertainers"[7] (who have for the most part replaced the local women), Korean bar owners, and local taxi drivers are the only civilians that they ever got to meet, and getting into fistfights with cab drivers, Jay bragged, had become almost a competition for some of his comrades, who tended to have run-ins with the typically older, male Korean drivers. The language of communication in Tongduch'ŏn was a mix of broken Korean and English, and Jay himself quickly learned how to say "Fuck off" and "I'll kill you" in Korean; "That's usually enough to drive guys away who wanna fuck with me," he added.

Finally, he brought up Ramirez, and gave me a description of the occurrence that reflected the extreme social and geographical distance that separates him and his soldier friends from the inner-city Korean student space of Hongdae:

> There was this guy who was charged with raping a 60-year old woman. I know the guy; he still claims he didn't do it. Well, I'm sure he came on to the woman, but … They were in one of *those* neighborhoods, you know. Where the only women you meet are prostitutes. But then, you know, the Korean media, they said that normal people are living in these areas, too. But of course, the soldiers, they don't see it that way. If you are in a certain neighborhood, you gotta be a hooker. That's the way they see it.

The crucial error underlying Jay's justification of Ramirez's actions—the woman may have been a prostitute after all—not only implies that violating a sex worker somehow constitutes a lesser crime than the attack on a "decent" female. In the particular context of Seoul, it also points to a gross misreading of a complex social urban space that Jay, with his limited knowledge of South Korea, is unable to fully grasp. Hongdae, in fact, is not one of "*those* neighborhoods" where sex is for sale; rather, it is an entertainment area popular with young Korean adults, in which, as a Korean friend of mine once put it, on your typical Saturday night out you have to "hunt for sex" rather than buy it. Jay's superficial knowledge of Hongdae—an area which he had visited only once—resulted in his conflation of the red-light districts near remote US military bases with this lively inner-city entertainment area mainly frequented by Korean students, artists, and unruly youth.

What is perhaps more interesting than his ignorance on the matter, though, is that Jay is embedded in a structure that *allowed* him not to care all that much whether the student district of Hongdae was, or was not, one of "*those* neighborhoods" where sex is for sale. His idea that any Korean woman he came across in "a certain neighborhood" necessarily needed to be sexually available to his comrades speaks of a certain kind of

Figure 1.2   Chongno entertainment district in downtown Seoul

dis-location of decades of GI experiences and behavioral patterns in Korea into the unknown territory of an experimental Korean student neighborhood in Seoul. It also hints at the gendered power relations in which this (mis-)understanding is embedded, structures of power which have—incidentally—come under heavy contestation over the last few decades.

An Anthropology of Militarism

The Korean peninsula today is one of the most heavily militarized regions on the planet, where the armed face-off between the northern and southern half has now entered its 66th year. At the end of the Korean War (1950–53), an armistice was signed—an old ceasefire that is broken at regular intervals when smaller fights erupt at land or sea between the contestants. The lack of a real peace treaty between the opponents has resulted in a permanent lock-down along the dividing line ironically named the Demilitarized Zone (DMZ). Once described by Bill Clinton as "the scariest place on earth" (Havely 2003), militarization around this particular border has reached such intensity that it has turned the buffer zone into the most heavily fortified space on this planet.

The Korean People's Army today consists of over 1.19 million soldiers, with an additional 7.7 million people in the reserve, which makes it the fifth largest armed force in the world. About 70 percent of North Korea's troops are stationed in close proximity to the border with South Korea (Bermudez 2001: 1ff). The South Korean Armed Forces, situated on the other side of the DMZ, currently have around 655,000 people as standing troops and another 3 million in the reserve, with a majority stationed in this border region as well. For the year 2007, it has been estimated that more than 30 percent (about $8 billion)[8] of the Democratic People's Republic of Korea's gross national income went into the defense sector, at a time when South Korea, with its $26.3 billion defense budget, actually spent a sum three times larger than its opponent (Moon and Lee 2010).

In the midst of such incredibly large local troop contingents, and the dispensing of such huge financial resources, which together have led to the ever increasing militarization of the peninsula, the number of US soldiers deployed in South Korea, which currently hovers around 30,000, may seem rather inconsequential. However, the continued presence of US troops in the South is of huge symbolic significance, pointing to the vast breadth and depth of US political, economic, and military engagement in Korea since the 1950s.[9] What is more, US bases in South Korea do not stand in isolation, but function as vital spatial nodes of geopolitics and US empire-making in the way they are connected to other US military installations worldwide.

As Catherine Lutz, in her seminal work on the topic, *The Bases of Empire*, points out: the "global omnipresence and unparalleled lethality of the US military, and the ambition with which it is being deployed around the world" are unprecedented in human history (2009a: 1). In this particular universe the United States has created, 190,000 US troops are joined by an additional 115,000 civilian employees, who populate 909 military bases worldwide. In 46 countries and territories, the US military has 26,000 buildings and structures valued at $146 billion to its name (Lutz 2009a: 1). "These official numbers," Lutz claims:

are entirely misleading as to the scale of US overseas military basing, however, excluding as they do the massive building and troop presence in Iraq and Afghanistan over the last many years, as well as secret or unacknowledged facilities in Israel, Kuwait, the Philippines and many other places. (2009a: 1)

The network that the armed forces of the United States have spun around the globe is truly staggering, certainly providing an ample field for potential research. During the last decade, there has been a growing interest in militarism and the soldier as a subject for a critical anthropology, with the US Armed Forces, in particular, coming into sharp focus in a number of articles and books (see, for instance, Baca 2010; Forte 2011; Gutmann and Lutz 2010; Lutz 2001, 2002a, 2002b, 2006, 2009a, 2009b; Network of Concerned Anthropologists 2009). The tremendous global presence of the US military and of its soldier representatives on the ground is not exactly a recent manifestation, however, so one may wonder why the US Armed Forces have only become an area of anthropological research as of late. To be sure, the current interest was partially sparked by the wars in the Middle East that the United States launched in 2001 and 2003. The large-scale mobilization of many sectors of US society in order to pursue the "War on Terror," as it were, did not come to a sudden halt at the doors of academia. In the wake of the US military engagement in Iraq, the Pentagon sought to actively recruit anthropologists into its war efforts via the Human Terrain System (Forte 2011; Gill 2007; González 2009; Schober 2010)—a recently deactivated (Jaschik 2015) program of the US Army that employed social scientists to provide cultural and social insights about the populations to be conquered. As a response to such massive cooptation attempts,[10] calls have been made by a number of anthropologists for researchers to turn their gaze onto the military instead (Gusterson 2007), a task which is to contribute to a larger investigation into the workings of a US empire sustained by its global network of military bases (Johnson 2004). The ensuing ethnographies, it was argued by Catherine Lutz (2006), would be complementary to more systemic writings on empire, as ethnographies have the potential to "question the singular thingness that the term *empire* suggests by identifying the many fissures, contradictions, historical particularities, and shifts in imperial processes" (2006: 593).

Very few anthropologists working on military issues seem to have sought to define the key term of "militarism" as a concept, a phenomenon in which the subject of the soldier is vitally embedded. Other social scientists have been more precise in their usage of the term, however (for a review, see Stavrianakis and Selby 2012). Historian Alfred Vagts, for instance, in an early definition from 1937 in his book *A History of Militarism*, points to an important facet of militarism in the way it "ranks military institutions and ways above the prevailing attitudes of civilian life and carries the military mentality into the civilian sphere" (1937: 11). Sociologist Michael Mann, in his *Incoherent Empire* (2003), speaks of militarism as "a set of attitudes and social practices which regard war and the preparation for war as a normal and desirable social activity" (2003: 16f). Feminist writer Cynthia Enloe, on the other hand, argues in *Does Khaki Become You?* that:

> militarization can be defined as a process with both a material and an ideological dimension. In the material sense it encompasses the gradual encroachment of the military institution into the civilian arena. [...] The ideological dimension [...] is the degree to which such developments are acceptable to the populace, and become seen as a "common-sense" solution to civil problems. (1983: 9f)

Perhaps the most expansive definition to date comes from sociologist Martin Shaw, however, who argues that:

> the core meaning of "militarism" should be specified not in terms of how military practices are regarded, but how they influence social relations in general. [...] Militarism denotes the penetration of social relations in general by military relations; in militarisation, militarism is extended, in demilitarisation, it contracts. (2012: 20)

While all the definitions above point to militarism as a process that involves an encroachment and expansion of the military into civilian terrain, Shaw, in particular, puts the emphasis less on discourse or ideology, but instead focuses squarely on social practices. Indeed, such an emphasis on *practices* lends itself to anthropological inquiries, and is crucial for my own understanding of the phenomenon as it may allow us to also make sense of the rather distinct situation in South Korea.[11]

GI Crimes and the Public Imagination

Images of US soldiers continue to haunt modern Korea. In the Democratic People's Republic of Korea (DPRK), depictions of long-nosed, villainous

American troops (often pictured in the act of torturing and murdering Korean women and children) serve as one of the stock characters in the state's manifold propaganda repertoire (Myers 2010: 131ff). While it may not come as a surprise that a sworn enemy of the United States would make use of such depictions, in the allied nation of South Korea, too, images of US troops as offenders and criminals can easily be found. In the Republic of Korea (ROK), however, it is not the state that functions as the main disseminator of such images, but civilian actors hailing from a leftist-nationalist spectrum. A brief glance at popular South Korean movies released during the last decade, for instance, will reveal a number of films made by progressive film directors that have at times been labeled "anti-American" (Ryan 2012) for their depiction of the US military presence in the country.

In the wildly popular film *The Host* (*Koemul*, 2006), for instance, an actual event, when an employee of the US Armed Forces dumped a large amount of formaldehyde down the drain,[12] is taken as the movie's starting point. In this fictional world, the chemicals have now caused the rise of a monster living in Seoul's Han river. In *Welcome to Dongmagkol*, a movie released in 2005, US troops are seen attempting to bomb a secluded, peaceful village miraculously left untouched by the Korean War, where a renegade team of North and South Korean soldiers join forces to prevent this mass murder at the hands of Americans. And *The Case of It'aewŏn Homicide* (*It'aewŏn Sarinsakŏn*), which attracted a sizable audience in 2009, is a movie based on an infamous murder of a Korean college student in Seoul's It'aewŏn neighborhood, with two Americans as the prime suspects of the crime. Faced with such images depicting murder, misconduct, and lawlessness surrounding the US military, one impression inevitably takes shape: in South Korea's popular imagination, too, the contentious figure of the violent US soldier will not go away.

To be sure, images such as these—representations of some of the negative aspects resulting from the complex encounter between US troops and the South Korean population—are only one part of a larger story I wish to tell here. I set out for Seoul in September 2007 on what would become a 21-month-long journey with the idea in mind of finding out more about both popular imaginaries about GIs *and* the actual encounters between US military personnel and locals. My main motivation in going to the capital of the Republic of Korea was one curious puzzle that I wanted to look more deeply into: South Korea was for a long time known as possibly the most US-friendly nation in the world, with the Republic of Korea being, as Bruce Cumings once put it, "one of the few countries that never said 'Yankee go home'" (2005:102). But over the last few decades, South Koreans seem to have had a drastic change of heart.

On December 14, 2002, for instance, an estimated 300,000 people attended candlelight vigils across the country to protest the death of two 13-year-old schoolgirls who had been run over by a US military vehicle (Cho 2013; Min 2002). In 2006, violent clashes erupted between farmers and activists, who faced thousands of Korean riot police when their rice fields in the village of Taechuri (near P'yŏngt'aek) were seized for the expansion of a nearby US base (Yeo 2006). Two years later, in 2008, during another round of candlelight rallies that erupted in Seoul, hundreds of thousands of protesters attended a series of protests after a ban on US beef imports was lifted, with anti-American sentiments running high once more (Lee J. 2012; Lee S. et al. 2010). And finally, over the last few years, the completion of a Korean naval base on Cheju Island has been delayed due to a number of protests. The opponents of this project argue that the US military will also have access to this ROK Navy-run facility, which may turn it into a key outpost for American attempts to keep maritime hegemony in the region intact (Kirk 2013; see also Pae 2014). These are just a few instances of recent public anger in South Korea over issues pertaining to the United States and its entanglement with the fate of the Korean peninsula.

Within this heated context, "GI crimes" (*migun pŏmjoe*) were repeatedly taken up by actors of the nationalist left as examples of the quasi-colonial nature of the long-term alliance between the United States and Korea. The National Campaign for Eradication of Crimes by U.S. Troops in Korea, for instance, a non-governmental organization (NGO) founded in the early 1990s that is opposed to US bases estimates that tens of thousands of crimes were committed by US soldiers against Korean citizens,[13] as approximately 1,100 to 2,300 crime cases involving US servicemen were reported annually between 1976 and 1991 (Moon 2010a: 354). Clearly the issue of violent soldier behavior, with those living and working in or near US entertainment areas predominantly affected, has often been a weighty matter of concern, made worse by the fact that addressing it publicly could very well land a person in jail until the years of the military dictatorship (1961–88) came to an end.

There was one particularly heinous offence, I was to learn, that over the years would become viewed as the quintessential "GI crime": the gruesome murder of a young Korean prostitute by the name of Yun Kŭm-i who was killed by Private Kenneth Markle on October 28, 1992—an event that, to this day, seems to represent people's imaginations in South Korea of what US soldiers are potentially capable of. In the months and years after the murder, the incident was turned into a central symbol of US domination within left-wing narratives; and the controversy caused years later by the rape case in Hongdae also needs to be read in light of the massive fallout caused by this brutal murder in 1992. At that time, as we shall see, the death of Mrs. Yun served as a starting point for widespread public agitations that would reappear with each new transgression of US military personnel or their dependents.

Undoubtedly, public outrage over violence committed by GIs against local civilians is not a scenario that is unique to South Korea. Wherever US troops have been stationed on foreign soil, controversies over the unlawful behavior of some soldiers have followed, with rape and murder cases, in particular, often sparking great shock among the affected local populations. Images of tortured Iraqi prisoners and of smiling US soldiers dishing out unlawful punishment in the Abu Ghraib prison spread throughout the world in 2003, and became emblematic of the larger injustice that the 2003–11 war in Iraq represented to many. Perhaps less well known is how individual crimes that were committed in the context of the peacetime stationing of US soldiers overseas have also frequently been turned into symbols of greater grievances by local anti-base movements. On September 4, 1995, for instance, three US servicemen stationed at Camp Hansen (Okinawa, Japan) kidnapped and raped a 12-year-old girl whom they encountered walking by the side of the road, an incident which led to large-scale anti-US military protests on this Japanese island (Angst 1995, 2001). In a similar, much publicized case in late 2005, a Filipina woman accused four US marines of gang raping her at the Subic Bay Freeport, with Lance Corporal Daniel Smith found guilty of rape a year later. Smith was eventually released from prison in 2009, after the accuser withdrew her statement, possibly in exchange for a US green card (Lacsamana 2011; Winter 2011). The disputed death of Alexander Ivanov, a Kyrgyz truck driver killed on December 6, 2006 by 20-year-old US airman Zachary Hatfield during a security inspection, led to many discussions about the Status of Forces Agreement under which US troops operated in Kyrgyzstan at that time (Cooley 2008). And, most recently, the death of Jennifer Laude, a Filipina trans woman, who in late 2014 was found murdered in a motel that she had entered together with US sailor Joseph Scott Pemperton, has put strains on the renewed military alliance between the United States and the Philippines (Talusan 2015).

In all these instances, violent acts committed by US servicemen against local civilians have altered the circumstances under which US American military personnel operate abroad. The South Korean case, however, is perhaps distinct in that the controversies over GI crimes had a rather limited impact on the actual security alliance, but still led to an immense loss of support among the civilian population. Additionally, these prolonged struggles over the US military presence in South Korea that were fought out over the terrain of violent incidents have had an unmistakable urban component that was largely missing from similar cases elsewhere, which is a point I shall turn to next.

Figure 1.3   Partial panorama view over Seoul

The Urban Setting of Seoul

The story I wish to tell here about the presence of US troops in Korea is not only about public sentiments over (gendered) violence committed by US soldiers, but also one in which the city of Seoul plays a rather central role: South Korea's capital has about 14 million people living within the actual city boundaries; Seoul's satellite cities in the Kyŏnggi province that surrounds the capital, however, seamlessly blend into this core metropolis, thereby forming a huge urban field (Friedmann and Miller 1965) marked by high infrastructural integration and extreme polycentricism.[14] The population of this Greater Seoul area, as of 2015, stands at 25,144,000 people,[15] which makes Seoul one of the largest urban areas in the world. In Mike Davis's terminology (2006: 5), this qualifies Seoul for the moniker of "hypercity" (that is, 20 million inhabitants or more), a kind of urban space that puts the "megacity" with its 10+ million residents to shame. Today, more than half of the population of South Korea lives in this Greater Seoul area, where population density is eight times higher than that of Rome (*Hankyoreh* 2009).

In addition to the nearly 30,000 soldiers permanently stationed in the country these days, the US military on occasion brings in hundreds of additional troops during military exercises. Over two-thirds of the US military bases inside South Korea are located within or close to the Greater Seoul area (Moon 2010a: 348). In this massive urban context, the number of military employees amounts to such a minuscule presence that US soldiers more resemble the proverbial needle in the haystack rather than the sizable troop contingent they would be in a less populous city. To complicate matters further, US soldiers are no longer staying put during their free time in the particular entertainment zones nearby US bases (kijich'on) catering to their needs. As one unintended consequence of rapid infrastructural development in the Greater Seoul area, previously significant distances between peripheral towns and inner-city areas have shrunk into manageable (albeit hefty) commutes. This is one explanation for why inner-city entertainment areas such as Hongdae have grown increasingly popular among American soldiers since the mid 2000s: these adult entertainment zones have simply come within easier reach through the extension of Seoul's public transportation network. The influx of GIs that this has caused has led to a number of symbolic dislocations affecting the particular inner-city entertainment district of Hongdae. This can perhaps be most acutely observed in the appropriation of the term *yanggongju* ("Western Princess," a derogatory expression for women working in camptowns), which was now used in order to label young female partiers who got mixed up with foreigners in Hongdae, a discussion that will be explored in chapter 6.

In reaction to the increased mobility of American soldiers in the midst of such a vast field site (that is, the Greater Seoul area), I opted for a research strategy that involved seeking out US servicemen and the disparate groups of people they encounter in various neighborhoods in and near Seoul. In addition, as my fieldwork progressed I also became more interested in those citizens of the area who had barely ever (or never) come into contact with US soldiers, but who seemed to be acutely aware of their presence in the city. So many of the Seoulites I met had never personally encountered US military personnel up close, but they still had many striking things to say about them, which is why "imaginaries," that is, a mediated kind of knowledge not based on first-hand experiences, became crucial to my understanding of US–Korean relations. Indeed, face-to-face encounters with GIs were by no means a prerequisite for holding rather strong opinions about them, I was to learn.

Beyond the mediated aspects of the at times fractious relationship between the US and South Korea, another goal of mine was to learn more about how people *do* engage directly with US soldiers when they encounter them in the vast urban terrain of Seoul. The widespread focus on violent acts, a frame[16] first utilized by the nationalist left, who sought to criticize the existence of US bases in the country from this particular angle, necessarily supersedes alternative stories of how civilians in South Korea come to encounter US military personnel in their daily lives. The actual meetings between soldiers and ordinary people that occur every day and night in Seoul are as manifold as the vast array of actors that engage in them: extraordinary or banal, antagonistic or smooth, erotic or dull, orderly or unrestrained, fleeting or lengthy, the potential is nearly endless. What the stories that will be told here have in common, however, is that the violent imaginaries that surround the US military presence in the country do function as the very backdrop, the imagined terrain, so to speak, against which actual relationships between US soldiers and civilians do take shape. And while the more hostile environment that GIs are stepping into these days has not necessarily narrowed the script of everyday possibilities, rarely did the people I spoke to entirely forget how contentious the US military presence in the country had become over the last few decades.

Adult entertainment areas as material spaces are crucial to my analysis as the sites for these encounters, as they are *the* prime locations where South Koreans and US military personnel actually do meet. Not only are inner-city entertainment districts (such as Hongdae or It'aewŏn), or more remote red-light districts (such as Tongduch'ŏn, Songt'an, or Anjŏng-ri) the locations in which a number of "critical events" (Das 1997) took place that pertain to the US military's contentious public standing in the country. They are also, significantly, the first terrains in which South Koreans can actually replace thoroughly mediated imaginaries about US soldiers with actual, first-hand encounters. And as I learned more about the historical differences, social complexities, and local specificities of these areas, I came to attend to the Greater Seoul area as an urban space made up of particular neighborhoods in which the US military's influence cracks, fragments, and is often altered into something new that is very much bound to the particular location I found myself in. In short, my journey also became a project about how places such as these entertainment districts are constituted, shaped and altered by those who find themselves in them, and how soldiers, local and foreign entertainers, Korean students and business owners, NGO people and anti-base activists all hold rather different stakes in them.

Violent Imaginaries as Social Practice

In *Base Encounters*, I will attend to the consequences of a particular, by now largely historical problem: how and why did South Korea go from being a country that was known as one of the most US-friendly in the world to one where the US military presence has become largely contested? Actors from the nationalist left, I was to find, have strategically utilized negative images of US soldiers as a counter-hegemonic discourse and a popular frame that later trickled into other sectors of Korean society. The details of this contingent process, which turned unfavorable depictions of US soldiers into a key component of a political project that allowed the positioning of the United States within a long historical line of intruders, will be further laid out in chapter 2. Before attending to the historical particularities, however, it seems appropriate to say a few more words about a theoretical notion that I call "violent imaginaries," which I use in order to make sense of this change of heart and its lingering after-effects.

Based on the historical and ethnographic context to be deciphered here, *violent imaginaries* will refer to *the social practice that describes how people make sense of US militarism through the reconfiguration of individual acts of violence into a matter that pertains to the nation*. In utilizing this term in such a way, I do not wish to make a claim to fictitiousness and call into question the widely distributed crime statistics published by Korean NGOs that point to rampant GI crime. Instead, I want to highlight that (a) violent imaginaries, in their materialization during a particular moment in history, constitute an *action* that is aimed at political change;[17] (b) much *mediation* goes into conjuring up such negative images of US soldiers, as most Koreans will rely on information provided by others rather than first-hand experience; and that (c) such depictions have become integral to a *nationalist frame* through which US–Korea relations are assessed. Such a definition, I believe, requires some additional clarification by way of a brief literature review. What do anthropologists actually mean when they say "imagination"/ "imaginary", and how is this connected to another, equally ubiquitous term, that of "violence"?

"Imagination" was for a long time a phrase that was largely limited to the field of philosophy. The preoccupation of philosophers with the complex relationship between reality and imagination has been traced as an idea from its very beginnings in the works of Plato and Aristotle, through medieval times and into the Renaissance in J.M. Cocking's *Imagination: A Study in the History of Ideas* (1991). In 20th-century philosophy, the term has been picked up by Jean-Paul Sartre, among others, who in his book *The Imaginary: A Phenomenological Psychology of the Imagination* (2004 [1940]), sought to solve long-standing philosophical questions as to the nature of human consciousness by utilizing this term. And, in recent years, Canadian philosopher Charles Taylor has written much on *Modern Social Imaginaries* (2003), which establish the kind of multiple modernities he is interested in, with social imaginaries standing for "a commonly shared understanding of how things go, as well as how things *should* go, in the collective life of a community" (Williamson 2004).

Taylor, like many other scholars working with the twin-terms of "imaginary" and "imagination," took many cues for his own work from the doyen of nationalism studies, Benedict Anderson. Anderson's magisterial book *Imagined Communities* was first published in 1983, the same year that also saw the appearance of two other classics on the subject: Hobsbawm and Ranger's *The Invention of Tradition*, and Gellner's *Nations and Nationalism*. While liberal philosopher Ernest Gellner showed that nations were entirely modern constructions, with nationalism in his view "primarily a political principle, which holds that the political and national unit should be congruent" (1983: 1), Marxist historian Eric Hobsbawm concentrated on how such modern nationalist movements were in fact excellent at fabricating myths and histories that served their own agendas.

Benedict Anderson, instead of focusing on concoction and fabrication, put his own emphasis on the question as to why nationalist movements are so effective at creating popular support. Anderson found one answer in the potency of communities not actually based on face-to-face encounters, but on thoroughly mediated and imaged belongings. Association via imagination, he argues, became ever more important with the onset of "print-capitalism," a period of modernity which allowed the reconfiguration of disparate populations into members of a nation as they were now becoming dimly aware of each other's existence. Thus, the nation is imagined, Anderson famously wrote, "because the members of even the smallest nation will never know most of their fellow-members, meet them, or even hear of them, yet in the minds of each lives the image of the communion" (1991 [1983]:6).

Anderson's intervention showed how nations can be built on such seemingly fleeting stuff as the collective imagination, and it was just a matter of time until anthropologists would also claim this terrain for their own investigations into various (trans-)national social phenomena. While the literature on imagination in anthropology is a vast and burgeoning field these days (for recent contributions, see for instance, Gibson 2014; Graeber 2015; McLean 2007; Salazar 2011, 2012; Severi 2015; Skinner and Theodossopoulos 2011; Strathern et al. 2006), the term is perhaps most firmly associated with the work of Arjun Appadurai, who turned "the imaginary" into a key phrase in his intellectual endeavor about how to make sense of globalization. In *Modernity at Large* (1996), Appadurai lays "the imagined" out as a potent field of practice that deserves more attention in our globalized day and age:

> The image, the imagined, the imaginary—these are all terms that direct us to something critical and new in global cultural processes: the imagination as a social practice. No longer mere fantasy (opium for the masses whose real work is somewhere else), no longer simple escape (from a world defined principally by more concrete purposes and structures), no longer elite pastime (thus not relevant to the lives of ordinary people), and no longer mere contemplation (irrelevant for new forms of desire and subjectivity), the imagination has become an organized field of social practices, a form of work (in the sense of both labor and culturally organized practice), and a form of negotiation between sites of agency (individuals) and globally defined fields of possibility. (1996: 31)

Appadurai further argues that the realm of the imagination may also help us to link "the play of pastiche (in some settings) to the terror and coercion of states and their competitors" (1996: 31). In his own work, however, Appadurai largely stayed clear of violent issues such as terror and coercion, and focused on more tranquil terrains opened up by his global ethno-, techno-, finance-, and other scapes. Given these choices, Appadurai may not be the best of all contenders to help us answer a number of questions related to violence and the US military, such as the following: How we can make sense of violent images that have been pertinent to shaping a social movement against US bases in South Korea? And why are negative depictions of US soldiers as violent (sex-)offenders on the loose so widespread in a country that is also an ally of the United States and heavily dependent on its economic, military, and social contributions?

Anthropologist David Graeber, in his article "Dead Zones of the Imagination" (2012), makes a more substantial attempt to analytically bring violence and the imagination together. In the broader literature on violence in anthropology,[18] Graeber notes that much attention has been paid to a kind of "poetics of violence" (for example, Caton 1999; Whitehead 2004: 55ff), that is, the focus has been on how violence is a kind of language that is utilized in order to communicate. This is a perspective that Graeber rightly critiques:

> Yes, violent acts tend to have a communicative element. But this is true of any other form of human action as well. It strikes me that what is really important about violence is that it is perhaps the only form of human action that holds out even in the possibility of having social effects *without* being communicative. (2012: 116)

"Violence," Graeber further argues, "may well be the only form of human action by which it is possible to have relatively predictable effects on the action of a person about whom you understand nothing," which allows the perpetrator to cut through all the "subtle work of interpretation" that goes into most human relations (2012: 116).

Another salient point made in "Dead Zones of the Imagination" is that regimes of violence create "highly lopsided structures of the imagination" (2012: 119). Whoever has the upper hand in a relationship, Graeber claims, rarely needs to ponder on the motivations of the people they dominate. "Those on the bottom of a social ladder," on the other hand, "spend a great deal of time imagining the perspectives of, and genuinely caring about, those on the top" (2012: 119). Hence, while violence is a tool that allows the more powerful to get away without knowing much about the Other at all, it forces the weaker component in a relationship ever more deeply into guess-work to render the seemingly arbitrary decisions made at the top more meaningful.

As will be analyzed in detail in this book, individual moments of violence can occasionally be turned into grand metaphors for the larger power structures that have allowed these events to occur in the first place, a process that anthropologist Marshall Sahlins has termed "structural amplification" (2005). In the South Korean context, Graeber's notion of the imagination as a one-way street of meaning-making may help us understand the commotion around incidents such as the Yun murder or the Hongdae rape case. These and similar occurrences were turned into moments that spoke to Koreans of the greater stakes involved in the US–Korea alliance precisely because ordinary people in South Korea have had to invest so much into the daily functioning of this relationship, while most US citizens have been (and still are) blithely unaware of the contingencies created by the presence of US troops in Korea. Violent imaginaries, I believe, arose as a powerful social practice in South Korea out of the sudden disturbance caused by a particularly heinous crime (that is, the murder of Yun Kŭm-i); they were constituted as a framework through which US soldiers could be perceived once the interpretative work that routinely greased US–Korea relations during less turbulent days was suddenly interrupted. In the aftermath of the brutal Yun murder, the incessant guess-work that went into building the asymmetrical relationship between the USA and Korea seemed to become meaningless for a while, as the murder starkly laid bare the power structures underneath that had made the event possible. It was only this stoppage, caused by a particular event and the public responses it triggered, I contend, that allowed for US–Korea relations to eventually tip over into something new.

To be sure, the timing of these contestations over the collective imagination pertaining to the US also had to be right: the conflict, we will see, gained heat at a particular moment in South Korea's history, that is, in the midst of the stormy period of democratization in the early 1990s, which allowed the rescaling of these imaginaries into the realm of larger political struggles. During preceding decades of military rule in South Korea, benevolent notions held about GIs had gradually been undermined by a growing sense of anger over the virtual immunity provided to US military personnel through the Status of Forces Agreement (SOFA) between South Korea and the US. But, even more damaging than unequal legal frameworks that, as we shall see, may have implicitly encouraged social irresponsibility among GIs on the ground, proved to be the daily US *Realpolitik* surrounding Cold War Korea, during which the American allies repeatedly placed security concerns over those of democracy. In such a way, violent images involving individual US soldiers were forged into a deliberate tool to fight a war over grander matters; the small was amplified into the large, the individual reconfigured into the structural, all with the goal to muster further outrage for a growing nationalist movement to redress decades of putative oppression.

This rise of violent imaginaries has also had an interesting spatial ingredient to it that will be further delved into throughout the book. The actual spaces where encounters between GIs and local or foreign women usually do take place—that is, the entertainment areas close to US military bases—were labeled as spaces of domination for the first time after the infamous Yun murder occurred. Areas next to US bases, where GI bars and clubs are typically clustered, became imagined territories of national shame, where American hegemony seemingly touched ground in its most violent manifestation. Fueled by a number of critical events involving US soldiers, debates over these camptown areas quickly escalated to such a degree that the American ally could undergo a metamorphosis into yet another unwelcome intruder.

Undoubtedly, "violent imaginaries," cast in such a nationalist form, also entailed a sacrifice of nuances and gray zones, and the eradication of alternative narratives and visions concerning the contentious encounters between soldiers and civilians. Whatever did not fit into the tale of violence and exploitation was filtered out, and the consequences of such a loss of complexity in the way camptown areas and their inhabitants have been imagined are still evident today. For one, it means that the voices of foreign sex(ualized) workers in camptowns—who by now make up the vast majority of entertainers in these areas—are practically silenced, as they do not fit into nationalist understandings of what these contentious zones stand for. To be sure, even in the most remote and marginalized camptowns, murder, rape and other forms of violence are by no means a daily occurrence these days. During field research, in fact, I found that for the (mostly Filipina) women employed in the GI clubs, it is not their clients that they fear the most, but the costly suspension between different countries and legal regimes that is negatively affecting their lives as migrant workers in South Korea.

Soldiers and Contentious Sexual Encounters

Access to foreign women, writes feminist author Cynthia Enloe, is one of the unspoken perks that come with joining the US military:

> Without a sexualized "rest and recreation" (R&R) period, would the US military command be able to send young men off on long, often tedious sea voyages and ground maneuvers? Without myths of Asian women's compliant sexuality would many American men be able to sustain their own identities of themselves as manly enough to act as soldiers? (1992: 23)

Enloe's work on the US military (see, for instance, Enloe 1983, 1989, 2000) has inspired and informed legions of social scientists to start to think outside the box, and to conceptualize the military as something other than a mere natural fact. Countering a peculiar omission in social scientific research that is possibly related to the fact that national security institutions are still among the biggest financial donors to the American academe (sponsoring much research *for* the military, and very little research *on* the military), a number of (largely female) authors have placed the issue of gender and the military at the heart of their work (see, for instance, Cohn 2012; Elshtain 1987; Shigematsu and Camacho 2010; Sjoberg and Via 2010; Stiehm 1996; Witworth 2004). In their writings, they have sought to fill the gaps in our knowledge of the numerous effects that the US military's global expeditions, expansion, and extensions have had on the existences of millions of people living through and around them, with local women and their engagements with foreign soldiers, in particular, coming into sharp view.

In some of these feminist contributions, the asymmetrical encounter between male US soldiers and females employed in the sex industry has served as the most important vantage point to approach the explosive issue of US bases overseas (e.g. Enloe 1989; Hoehn and Moon 2010; Moon 1997; Sturdevant and Stoltzfus 1992). US imperialism, in this context, is seen as a project that is vitally held in place by a form of virulent masculinity that is enacted by soldiers in the everyday contact zones near US military installations with the aim of dominating the local population via the bodies of women. Out of all institutions, the military is most closely associated with the formation of hegemonic masculinity (Shefer and Mankayi 2007: 192), hence the entertainment areas that US troops in Seoul primarily inhabit in their free time are understood as realms that have been critical in the spawning of poisonous gender relations imbued with violence among nations allied to the United States.

The value of this pioneering work should not be underestimated; yet there is at times a certain tendency to eradicate nuances evident in this body of literature. It is implied in some of these writings, for instance, that military people necessarily need to be seen as the perpetrators in what has widely been called "militarized prostitution." Local brokers of prostitution who run the GI bars and clubs near US bases (often former sex workers themselves) do not figure in the picture very much, as the acknowledgment of their role could potentially complicate the underlying binary opposition of foreign military perpetrator on the one hand and local female victim on the other. Casting the women in the role of victim, however, is an analytical move that may run the risk of denying them the agency to manage their own lives and fortunes. The equating of all types of sexual encounters involving GIs with acts of violence also tends to level out differences and subtleties that can be found in the wide range of encounters that take place between civilians and soldiers in the urban space of Seoul. The blurring of boundaries between consensual sex and violence is, of course, not a unique feature of writings concerned with gender issues surrounding the US military, or, for that matter, of the discourses utilized by anti-base activists in South Korea. Rather, it is symptomatic of many scholarly debates on the subject of prostitution itself (on this discussion, see, for instance, Agustin 2007; Berman 2003; Doezma 1998; Kempadoo 2005; Kempadoo and Doezma 1998; Weitzer 2000, 2005).

For the purposes of this book, these discussions raise the question as to whether there are alternative ways available for us to make sense of US soldiers in relation to the women they become sexually involved with near US bases overseas. One possible way out of such binary victim–perpetrator scenarios is offered by literary scholar Lee Jin-kyung, who argues that it may be more useful to understand US soldiers as both agents and victims "of the state's necropolitical power" (Lee J. 2009: 656). Lee believes that the agentive power of soldiers is significantly held in check by the inflexible labor circumstances in which they find themselves while serving in the US military. Difficult labor conditions also inhibit the "entertainers," who need to play by the rules laid down by the clubs they work at. The women laboring at these entertainment facilities these days come from the most marginalized sectors of Korean society, or, more frequently, have been recruited through work agencies in the Philippines or the former Soviet Union (a topic that anthropologist Sealing Cheng has examined in her monograph *On the Move for Love*, published in 2010).

The soldiers, on the other hand, often hail from impoverished ethnic minority sectors in the United States, or other social strata of US society where access to higher education and better-paying jobs may not be readily available to them. Both the sex(ualized) workers servicing GIs and their lower ranking US military clients can thus perhaps be described as part of a globally mobile working class, who find themselves thrown together in a third location, that is, the (sub-)urban peripheries of Seoul, South Korea, where they have been brought in order to perform their work-related duties. In a provocative move, Lee Jin-kyung has called this parallel convocation of male and female labor in camptowns expressions of a "sexual proletarization," which "defines the process of mobilizing respectively gendered sexualities into various working-class service labors, such as military labor, military and industrial prostitution, and other sexualized service work" (2009: 656).

While the point is well taken that foreign entertainers in kijich'on may represent something like the quasi-natural counterpart of the American soldiers in the sense that these protagonists all hail from the lower ranks of the global workforce, the labor mobilization of female sex(ualized) workers in the end *does* involve a much more pronounced invocation of their sexualities than the mobilization of military labor. The specific expression of the soldiers' sexualities in the camptown areas, which is what in the end keeps the system of prostitution in these neighborhoods alive and well, is certainly fundamentally shaped by the constant emphasis on strength and virility during working hours inside the base. At the same time, the fact that soldiers seek out foreign entertainers is for the most part an unwanted byproduct of the work regimes they are subjected to, an unintended consequence, if you will, that is to be held in check by military authorities, rather than an integral part of the labor tasks performed, as it is the case for the women in camptown. In a similar manner, sociologist Linda McDowell, in her book *Working Bodies* emphasizes that for men, "the main attribute associated with the masculine body at work *is not its sexuality* but its strength" (2009: 129, emphasis added). Masculinity, she further claims, is a highly class-specific matter, with men hailing from working-class backgrounds often finding that the strength of their bodies is indeed the only commodity they can viably sell in a rapidly de-industrializing nation like the US. "Sports and the army," McDowell says, "are often escape routes for young men brought up in relatively deprived circumstances with few educational credentials and ever fewer options in the feminized bottom of the service sector labour markets" (2009: 130).

The virility of the soldier, however, is certainly not a static characteristic, but rather, like so many expressions of gender, may best be understood as "a dynamic and emergent property of situated interaction and in need of ongoing accomplishment" (Pyke 1996: 528). If these forms of masculinity indeed only arise through effort and interaction, then camptowns near US Armed Forces installations serve as crucial terrains outside of the world of the military base in which various forms of masculinities and sexualities are (con)tested, shaped, and occasionally subverted. This is what makes *encounters* such a fascinating terrain to study in the context I have sought to make sense of: while shaped by the larger structural forces that have brought US troops to reside on a permanent basis on Korean soil, at the same time the specific meetings between individuals outside these facilities always include an instant of open-endedness, ambiguity, and surprise. Some agentic moments in the encounter between soldiers and sex(ualized) workers, then, may also emerge from their acknowledgment of their similarities, and from their attempts to shape romantic or other alliances in spite of a system that only encourages the commodification of their encounters through the exchange of sexual services for dollars.

The Structure of This Book

Before delving into the world of US camptowns in South Korea, in the chapters to follow, I will deploy an analytical perspective on Korea's history that foregrounds the critical conjunctures between politics, the economy, and the social. Looking at the crucial moments that have inaugurated the drastic change of heart that saw South Korea move from being one of the most US-friendly countries in the world to a place fraught with controversies over the American military presence, the perplexing trajectory of the national question in Korea will come to the center. Starting with the country's early encounters with colonial projects that eventually forced Korea to become part of the capitalist world system, the reader will be guided up to the moment of division after World War II, and to the eruption of armed conflict that culminated in the Korean War (1950–53).

The repressive Park Chung-hee era that followed in the 1960s gave rise to what I call "capitalism of the barracks": South Korea's double strategy of massive labor exploitation at home, and the interweaving of big business with the US military build-up in Korea and abroad. It was during this time, also, that the gradual realization among ordinary South Koreans would first emerge that the USA was not going to help facilitate the full package of democracy that it had promised the local population since its involvement on the Korean peninsula began. The most significant moment that led to the accelerated dissemination of such sentiments came in 1980, when the Korean military dictatorship sent troops against protesters in the city of Kwangju, which left hundreds of people dead. After Kwangju, key leftist actors would increasingly express their anger over the United States' involvement with a series of regimes that ruled the country with an iron fist until the early 1990s.

Chapter 3 will be dedicated to an account of the camptowns (or kijich'on) next to US military bases, and the social and imaginative role they have come to play in South Korea's post-war history. The Yun murder and its ramifications will be laid out in detail, with special attention paid to the nationalist discourses that emerged around the event. In particular, I will explore the social and economic factors that have shaped Tongduch'ŏn's ville. This small GI entertainment town, where the Yun murder took place, has always figured centrally in the imagination of some writers of the so-called *minjung* democratization movement. Actors from the left had set themselves the goal of wresting the definitional power over the nation out of the hands of the generals who had been in charge of the country's fate for so long, with a kind of camptown fiction—a popular genre of writing focusing on the lurid details of daily sexual exploitation in these neighborhoods—being utilized among disenfranchised, predominantly male leftist authors during the 1970s and 1980s. In an attempt to make sense of such imaginaries, anthropologist Marshall Sahlins' notion of structural amplification will be applied to explain how the GI red-light districts in the early 1990s came to be understood as the very locations where US domination touches ground. The prostitutes employed in these spaces, I shall argue, were shaped into symbols of a ravaged nation, and their actual experiences were silenced in favor of a streamlined nationalist narrative that proved easy to think and act with in the midst of a crisis over sovereignty.

Building on these insights, in chapter 4 I will tell the stories of some of the women living and working in these spaces today. The impoverished South Korean females of earlier days have by now largely been replaced by Filipina and Russian entertainers, who dream of a ticket to America as much as their Korean predecessors did, and many of whom, in the meantime, squeeze a meager living out of prostitution in these deeply claustrophobic sex-scapes. I shall focus on the strategic and romantic alliances shaped between women and soldiers by utilizing Sandya Hewamanne's (2013) notion of "preoccupation," which I believe captures well the intense emotional and sexual involvement of the women with US soldiers in the villes. While violent escalations occasionally affect the women's lives, the focus will be placed on the fearful suspension between different countries that the entertainers experience. Their uncertainties are heightened by the deeply precarious working conditions and the strict visa regimes that they must submit themselves to, and that add further pressure to their already vulnerable state as transnational female workers in South Korea.

In chapter 5, we will leave these peripheral spaces behind and move into the capital terrain. It'aewŏn is an entertainment area that is adjacent to the largest US military base in the country, the Yongsan US Army garrison in central Seoul. In this urban setting, the US military presence has had many unintended consequences that have turned the neighborhood into an unlikely incubator for social (ex-)changes. While during the military dictatorship a local containment strategy of sorts was at work that sought to keep US influences under control and within the boundaries of the "special district," nowadays It'aewŏn attracts impossibly diverse crowds outside of the military orbit. The arrival of sexual and ethnic minorities in the neighborhood has completely changed the urban landscape of the district. I will look at a distinct ambiguity that characterizes the neighborhood today: the uneasy positioning of the area between allure and repulsion which seems to dominate many people's imaginaries, and which is a phenomenon that I shall call It'aewŏn *suspense*. The at times very rowdy practices that take place in It'aewŏn prove to be both dangerous and creative, and have simultaneously engendered the destruction and production of social meaning and order. It'aewŏn's freedom—the counterintuitive liberties that some groups such as homo- and transsexuals, Muslims and other migrant communities have found next to the base— arises out of a suspension of the area between competing sovereignties, which have turned it into a loaded transnational terrain where images of coercion and persuasion reign with equal force.

In chapter 6, I shall finally focus on Hongdae, an inner-city student neighborhood that has ascended to nationwide fame ever since the early 1990s, attracting economically and politically disenfranchised groups such as Korean students, artists, and other rebellious young people. The progressive mixture to be found in Hongdae has also drawn in quite a few US soldiers, alongside other foreigners who have come to frequent the bars, clubs, and outdoor areas of this neighborhood. While GIs enter the urban space of It'aewŏn on an equal or privileged footing compared to other entertainment seekers, they are much less welcome in Hongdae, as we shall see, where many clubs have in the past refused them entry.

Unlike in the camptowns, sex in Hongdae is usually not for sale, but rather, one has to "hunt for it." Due to the ever increasing numbers of foreigners "on the hunt" for Korean men and women, Hongdae has become a bone of contention for the wider public, with the controversies around the neighborhood at times spiraling into a veritable panic over the putative moral corrosion of young Koreans by outside influences. The figure of the "Western princess" (*yanggongju*)—an old, derogatory term used for the sex workers laboring at the GI clubs in the remote camptown areas that we encounter in chapters 3 and 4—makes an unlikely reappearance in the geographically distant space of Hongdae, where anxieties over national purity and contamination at times have loomed large.

Finally, zooming out again from these battles over the meaning of a Seoul entertainment district "contaminated" by the presence of male foreigners, in my conclusion I will summarize in broad brushstrokes the ways in which the contentious history of the camptowns has figured in the South Korea of today. While the days of unassailable US supremacy in the region may have ended for good, the violent legacies, perilous imaginaries, and ambivalent encounters that the US military presence on Korean terrain has given rise to will possibly besiege all parties involved for many more years to come.

2

Capitalism of the Barracks
Korea's Long March to
the 21st Century

Nation(s)-in-Arms

It has become something of a routine horror show by now, played out at regular intervals, albeit with a new set of actors involved these days. When North Korea threatened South Korea, Japan, and the United States with nuclear war in the early spring of 2013, the only recent addition to the familiar rhetoric of "sea of fire," "nuclear flames," and "all-out war" coming from the North was that these words were now uttered by a new leader, with the young Kim Jong Un following in the footsteps of his deceased father Kim Jong Il. The Western world, as always, was quite alarmed, with the previous round of conflicts on the Korean peninsula already long forgotten. In late 2010, North Korean artillery had targeted South Korea's Yŏnpyŏng Island in the midst of a joint US–Republic of Korea (ROK) military exercise. The island, located 12 kilometers from North Korean territory and close to the disputed maritime border between the two states, was shelled heavily for an hour, resulting in the deaths of two civilians and two soldiers on the Southern side. Earlier that year, the sinking of the South Korean Ch'ŏnan naval ship—blamed on a North Korean torpedo— brought 46 casualties; this was another event that sparked major debates about the possibility of a broader military escalation in Korea. In the midst of such skirmishes, the uncanny persistence and staggering complexity of this conflict, which is being handed down to new generations of actors, is rarely fully discussed.

In order to understand the seemingly never-ending crisis in Korea, we need to also investigate the continuous US military involvement on South Korean soil. That is, the complex history, the global context, and the possible trajectories of the security alliance between the United States and the Republic of Korea need to be explored—an alliance that has fundamentally shaped this young democracy in North East Asia that is still riddled with countless contradictions stemming from past authoritarian governments. In this chapter, specific conjunctures, that is, moments in history during which local and foreign political and economic interests aligned in significant ways, will be highlighted. While the main focus will be placed on the South, it will prove crucial to keep the North in mind as well, and to investigate the emergence, maintenance, and modification of an entire "division system" (Paik 2009) that the peninsula has been subjected to since World War II.

In the southern half of the peninsula, as we shall see, the close affiliation of the succession of anti-communist regimes in South Korea with the United States of America proved to be a double-edged sword. On the one hand, the United States and its troops on the ground played a vital role in financially, politically, and militarily sustaining the various authoritarian leaderships in Seoul. On the other hand, the Americans also brought a gradual exposure to some of the promises of Western liberalism along, which were rapidly soaked up by local actors from the left. The very American ideal that civic freedoms may be attained together with economic prosperity would play a crucial role in bringing an end to authoritarianism in South Korea. A prolonged exposure to such notions eventually also spelled doom for the widespread support of the US among the Korean population, as one glaring contradiction gradually became visible: during its lengthy engagement with East Asia in the 20th century, the US government was often found to be verbally espousing democracy while at the same time backing illiberal dictators in South Korea.

Before we can discuss further the impact that such historical inconsistencies have had on the ground, however, it will be necessary to go back to the very beginning of Korean modernity. Three central questions will guide us through our (necessarily condensed) investigation of South Korea's long march to the 21st century: How exactly did the two Koreas come to imagine themselves as combat-ready nation(s)? Why have North and South diverged so drastically in the particular paths that their leaders chose to take? And what role has the United States and its armed forces played in all of this?

"A Shrimp Amongst Whales" (1895–1960)

In South Korea, there is a popular saying that is frequently brought up when talk turns to its repeated historical subjugation to outside forces.

"Korea is like a shrimp amongst whales," I was told on occasion by people from all walks of life. Rhetorically characterizing their country as a small nation desperately trying to maneuver its way among the big players of the area, this proverbial piece of wisdom has occasionally been interpreted as a starting point for a regional power analysis (see, for instance, Shim 2009). However, I would prefer to treat such an image as a condensation of a popular understanding of the Korean nation. This widely favored phrase, I believe, contains an important evaluation of Korea's prolonged attempts to find its bearings in a rapidly changing world. More often than not, the accelerated changes Korea found itself thrown into from the mid-19th century onward have led to unfavorable consequences for its people. A sense of crisis over sovereignty that emerged during that particular time would subsequently drive actors both North and South of the 38th parallel into embracing ever more militarist understandings of the nation.

The notion of a failed national history is deeply embedded in such an imagination of the country as a helpless, small creature among big predators. This negative framework deployed for an understanding of Korea's problematic past is also often linked to the country's supposed "failure to establish an independent nation-state and to fully purge collaborators and tainted political and social leaders immediately after 1945" (Lee N. 2007: 3). The significant delay in achieving full national sovereignty—putatively caused by both imperial Japan and the super-power of the United States—would later become a leading narrative that mobilized the left-wing forces of the country (Ceuster 2002). Before the Korean peninsula was propelled onto its particular 20th-century path, which brought two foreign occupations, the country's division, a civil war, and several (military) dictatorships ruling the country, it was first to experience a deep internal crisis brought about by the arrival of Western imperial powers in the region. The violent encounter between East and West in the second half of the 19th century, which led to the inclusion of East Asia in the capitalist world economy, would trigger a prolonged state of emergency in Korea, a period of frenzy during which some factions of the country's elites for the first time sought to frame their increasingly desperate struggle for survival in national terms.

Japan's Rise, China's Decline

As Commodore Matthew C. Perry and his squadron of US Navy ships coerced Japan into opening up its shores to Western trade in 1853–54, the first few cannons pointed at the Japanese harbor town of Uraga inadvertently meant the end of an era for neighboring Korea as well. In conjunction with the First Opium War of 1839–42, which had already brought China to its knees, this instance of gunboat diplomacy would eventually lead to the overturning of the old political order that had organized relations between the three reclusive North East Asian kingdoms of China, Japan, and Korea. For Japan, the signing of the Convention of Kanagawa (1854) brought internal political and economic turmoil that culminated in a shift of power, bringing about the Meiji Period (1868–1912) (Pyle 1996). The centralized state that emerged as a consequence pushed through fast-paced political reform, accelerated industrialization, and rapid militarization. Japan's elites deliberately modeled their country after Western examples. Prussia-Germany's successes, in particular, were being emulated: the compressed modernization of the Central European country that had successfully managed to catch up with the pace-makers of France and Britain deeply impressed the Japanese, who sent some of their best students to learn from the Germans (Miyake 1996). Prussian military structures and techniques, in particular, were copied, bringing a type of militarization to East Asia that was tightly interlinked with a state-professed nationalism that proved to be very useful for Japanese purposes (Anderson 1991 [1983]: 94f).

Japan, in this way, became the first country in the region where local elites initiated a process of rapid modernization from above in response to the more technologically advanced international actors in the world system that they saw themselves threatened by. And soon enough, influential 19th-century Japanese authors such as Fukuzawa Yukichi, smitten by the glories of the West, were propagating that no time could be wasted "to wait for the enlightenment of our neighbors [China and Korea] so that we can work together toward the development of Asia" (quoted in Atkins 2010: 18). The notion of taking a shortcut on the proverbial civilizational ladder, a social Darwinist image that foreign military men, diplomats, and traders had brought with them during their exploratory missions into the region, would now quickly gain a following in Japan. And indeed, from a statement such as: "It is better for us to leave the ranks of Asian nations and cast our lot with civilized nations of the West" (quoted in Atkins 2010: 18), it was just a short step toward aspirations of bringing the new-found light of civilization to those "bad Asian friends" Japan saw itself surrounded by.

Pre-annexation Korea is synonymous with the Chosŏn dynasty that ruled the largely agrarian feudal state for many centuries (1392–1910). Korean society, historian Carter J. Eckert argues:

was controlled to the very end by a small aristocratic group of landed families [… i.e. the *yangban*] who were able to perpetuate an oligopoly of wealth and power by strategic marriage alliances and domination of the state examination system, through which important political posts were granted. (Eckert 2000: 3)

Given this feudal context, and with very little industry to speak of that would have facilitated the rise of a home-grown form of capitalism, Korean elites were caught almost totally unprepared in the latter half of the 19th century, when heavily armed Western powers came knocking on the doors of the countries of North East Asia.[1] Chosŏn Korea and its feudal leadership stood for an extreme isolationist politics, which, over the centuries, had turned strangers of non-Korean descent on Korean soil into an oddity beyond all measure. Its international commerce "was officially restricted to tributary trade with China and, to a lesser degree, with Japan" (Eckert 2000: 8). While the nearby Chinese empire incontestably functioned as the center of cultural and political gravity for much of pre-modern Korean history, even this closely affiliated "Middle Kingdom" was kept at bay by a Korean leadership that sought to curb interactions with the supreme power of the region to a bare minimum. China usually dispatched a few official missions per year to the Korean kingdom that it considered a vassal, yet beyond the tributary goods sent to Beijing to appease the powerful neighbor, no further exchanges took place (Cumings 1997: 90).

After the Second Opium War (1856–60), however, the power of previously almighty China was in decline, a fact that was driven home to Koreans by the Kanghwa treaty between Korea and Japan in 1876. Following an earlier appearance of a Japanese gunboat at the island of Kanghwa, Japan now coerced Korea, which had until then been protected from such aggressions by its vassal status in relation to China, into opening its ports to international trade and severing its last formal ties with the Middle Kingdom. Consequently, it was not a Western power, but Japan that would play the most crucial role in forcibly introducing Korea into a rapidly changing world over the next few decades—a fact that had far-reaching consequences, as the South Korean intellectual Paik Nak-chung rightly notes: "because the capitalist world-system imposed its colonial rule through an Asian surrogate rather than through direct rule by a Western state, its Eurocentrism worked more insidiously and in some senses more effectively" (2000).

Early Korean Nationalists

Together with smaller encroachments by the United States,[2] France,[3] and Russia,[4] the predatory reach of a newly invigorated Japan left its mark on the morale of Korea's population: an "atmosphere of 'unprecedented crisis', permeated with the fears of 'demise' and 'extinction', became a dominant element in the better-informed circles of the capital [Seoul]" (Tikhonov 2003: 82). Within such an environment of widespread fear, a loose network of intellectuals was to form the first nationalist movement of Korea (Schmid 2002). Appalled by such dramatic events as the Japanese-sponsored assassination of Korea's Queen Min in 1895,[5] which made it clear that the country's fragile sovereignty was indeed in grave danger, they went to work. Most of them were writing for a series of newly established print media outlets that sought to imagine, for the first time, what the nation of Korea could possibly mean to its subjects. In the long run "the knowledge produced by these individuals and groups established the basis of modern Korean nationalist discourse" (Schmid 2002: 3f).

Contradictorily, they saw themselves both in opposition to and inspired by the rise of Japanese hegemony that their country would soon fall prey to, in such a way indeed verifying claims made by Anderson (1991 [1983]), Gellner (1983), Hobsbawm (1991), and others that the emergence of modern nationalism is often more a sign of the inclusion of a country into the global world order than an effective form of resistance to it. The questions that these early Korean nationalist writers pondered proved to be similar to those that had been preoccupying Japanese writers, politicians, and bureaucrats over the preceding decades. Besides concerns over Japanese and Western powers' continuous attempts to infringe the country's sovereignty, questions of Korea's territorial boundaries and the (il)legitimacy of its current leadership, as well as issues pertaining to civilization were prominently discussed: "Seen as part of a new global ecumene, the [Korean] nation needed to be brought into narratives of world history that plotted the trajectory of all nations along the same lines, ultimately leading to the modern" (Schmid 2002: 7f).

The burning issue of Korea's place (and time) within the recently created universal trajectory of civilizational progress was fundamentally influenced by social Darwinist readings spreading from Japan that had gained much ground in Korea by then (Tikhonov 2003, 2010).[6] From this angle, questions of modernity and civilization were condensed into imaginations of the world's nations as engaged in a continuous vicious battle for dominance and survival. Within this scheme, Korea was understood to be ailing because of an intertwined lack of military prowess and manliness, with the connection between failed nationhood and failed manhood that had previously puzzled the Japanese now being tackled in Korea (Jager 2003). In particular, it was the yangban, the stereotypically intellectual aristocrat in power who was portrayed as too effeminate to deal with the tasks at hand, with a new militarized man to be put in his place (Jager 2003: 3ff).

Influential historian Sin Ch'ae-ho (1880–1936), for instance, insisted "on both [the] crucial importance of 'military spirit' for the 'fate of the nation', and [the] deleterious effects of Confucian literary education on Koreans' military prowess" (Tikhonov 2003: 94; see also Em 1999). Despite his thoroughly anti-Japanese position, Sin would hail a form of militarized masculinity he saw shaped in its ideal form in Japan, embracing this particular model as a vision for the new Korean society that he envisioned:

> Physical education exercises the body, strengthens the will, and, by practicing certain skills, develops soldiers. [...] There is no one in the entire country [of Japan] who does not go through this military training. Students are future soldiers and merchants former soldiers; machinists too are future soldiers while farmers are former soldiers. Only when a country can count on all of its people to become soldiers in time of mobilization can it be a strong nation. (quoted in Jager 2003: 7f)

In order to counter the progressive "co-optation of areas of [Korean] nationalist thought developed autonomously before the 1905 Protectorate" by Japanese imperialists (Schmid 2002: 15), Sin Ch'ae-ho proposed a cosmology of the Korean nation as an ethnic community—a concept that is of great significance up to the present (Shin G.W. 2006: 2). He promoted the adoption of a cosmology of the nation that stressed ethnic/racial unity, which he backed up with a long genealogy of the putatively ancient and uninterrupted bloodline of Koreans. This also entailed imagining a direct linkage between Sin's contemporaries and ancient Koreans reaching all the way back to Tan'gun, the mythical half-god who was supposed to have founded the first Korean kingdom in 2333 BC. With such an incorporation of mythical history into a conception of what the nation might be, buttressed by the incorporation of other social Darwinist elements into the ideology of Confucianism, the discursive grounds were laid for the emergence of Korea as a nation-in-arms.

Korea's Annexation by Japan

Following Korea's annexation by Japan in 1910, Sin Ch'ae-ho, like so many of his generation, went into exile in China, where he increasingly became involved in the anarchist movement.[7] This transformation from Darwinist nationalist to committed anarchist undergone by someone who was arguably Korea's most influential historian in the early 20th century exemplifies the kind of political turmoil that afflicted the anti-colonial movement of the country, with the eventual split into a left and a right national camp already emerging on the horizon. The beginning of Japanese colonialism had brought severe crackdowns on Korea's nationalists, many of whom, if they had not fled the country in time, were now imprisoned or put to death. After the first shock, however, nationalist forces would realign themselves, encouraged by both Lenin's appraisal of self-determination and Wilson's 14 points that tackled the same crucial question of national sovereignty from different ideological angles in 1918. In March 1919, mass demonstrations against the colonial power swept through much of Korea, leading to the deaths of thousands during the suppression that followed. Facing international condemnation after the brutal quelling of the uprising, the Japanese reacted by introducing a new "cultural policy," which meant that Korean publications and organizations that had been banned for many years could resurface again for a while (Cumings 1997: 156).

During the renewed frenzied activities undertaken by nationalist activists at this time, diverging solutions to the schizophrenia-inducing conundrum of nationalism versus imperialism that Korea faced were offered up. The rifts between those who saw themselves propagating a Wilson-type liberalism as opposed to those enchanted by a potential communist revolution à la Lenin gradually deepened. Temporarily suspended by new crackdowns on Korean nationalists by the colonial power in the late 1920s and 1930s, these conflicts exploded with much violence after the end of colonialism (Cumings 1997: 154ff). Interestingly, it was Sin Ch'ae-ho who would once again provide some of the terms that would dominate these new struggles. In his Chinese exile, he started to wholeheartedly embrace the notion of a transnational revolutionary utopia. His earlier ideal of the nationalist warrior-hero was now replaced with that of the global *minjung* (oppressed masses) that were to be lifted out of misery (Jager 2003: 14ff)—a concept that would gain much currency over the decades to come among the nationalist left in his home country.

The Japanese occupying forces, in the meantime, used their 35 years of imperial rule in Korea to dramatically reshape the face of the country. The capital Seoul, in particular, was subjected to much change, with its muddy, narrow streets turned into paved, well-lit boulevards lined by grand-scale buildings occupied by Japanese officials. This attempt to bring the city's physical landscape in line with that of Japan's own urban centers went hand in hand with another project that aimed at incorporating Korea's citizens into Japan's populace. The new powers-that-be sought to forcibly assimilate Koreans into their vision of nationhood,[8] with Koreans figuring in the Japanese racial equation as a somewhat inferior people, yet still one endowed with a capacity to integrate. Hence tactics such as the forced adoption of new Japanese names were deployed, and schooling was now mandatorily undertaken in the Japanese language—measures which the colonial forces thought would guarantee the assimilation of Koreans in due time (Caprio 2009: 200).

With rapid industrialization now under way, the peninsula's natural resources and its human capital were to be exploited for the maximum benefit of the Japanese archipelago. After the outbreak of the Asia-Pacific War, labor shortages in the industries of Japan meant that over 5 million Koreans were now shipped to Japan and other destinations in the empire— either voluntarily or by force—and put to work under often horrendous conditions. In this way, up to 2 million Koreans ended up in mainland Japan by the time the war had ended, with several hundred thousand Koreans also making their way to Manchukuo (Cumings 1997: 177f). Driven by the promise of splendid opportunities awaiting them in this puppet state set up by Japan in China's Manchuria after 1931, Koreans would often join the police corps and military units of the Japanese imperial regime to better their lot. Incidentally, some of South Korea's later political leaders were among those earning their first military experiences as part of the Japanese Armed Forces. At times, the Koreans incorporated into the imperial military would find themselves up against Korean guerrilla fighters (among them, most prominently, Kim Il Sung) who had chosen to join the other side, that is, they had become communist guerrillas who sought to liberate Manchuria from the grip of the Japanese (Han 2005).

The most infamous incorporation of labor into the Japanese imperial machine, however, was the targeting of young Korean women and girls for the benefit of the military, with their sexualities exploited in the so-called "comfort women" system. As early as 1932 and continuing into the 1940s, up to 200,000 females of different national backgrounds, the majority of whom were from lower-class Korean families, were deceitfully recruited or directly forced into prostitution for the benefit of the imperial troops. Placed in so-called comfort stations that were established practically everywhere in Japanese-occupied Asia, these divisions were either run by the military or placed in the hands of local business owners who were entrusted with the task of facilitating the repeated rape of these women, a vast majority of whom did not survive the war.[9] Partly due to these dreadful events, the rape of Korean women by foreign soldiers has gradually become one of the key tropes used by later generations of nationalist authors who sought strong images for the suffering of their nation under foreign rule. Within this logic of nationalist resistance, miscegenation—and the racial, cultural, and moral contamination it may lead to—was understood as the biggest threat to a small Korean nation under constant duress (Jager 2003: 72f), a reasoning that can be traced in fragments all the way up to the anti-US military movement of today.

End of Colonialism, Division of the Country

After the US nuclear attack on the cities of Hiroshima and Nagasaki in August 1945—which, in addition to causing the deaths of up to 200,000 Japanese civilians, also killed an estimated 30,000 Korean forced laborers employed in military industries there (Tong 1991)—Japan's defeat was quickly announced. With the Japanese expansionist dream having burned out for good, new hopes for independence arose among those who had been subjugated to Japanese rule over the previous few decades. But for Korea, this post-annexation moment soon proved to be only a short phase leading to yet another foreign acquisition of its territory: the country would now be split into two occupied zones, bringing the peninsula under US and Soviet military rule that would last for three years (1945–48).[10]

Six days after the bomb had dropped on Hiroshima, on August 15, 1945, two young US colonels, Dean Rusk and Charles Bonesteel, were given 30 minutes by their superiors to find a dividing line that would allow Korea to be split into two. Without any Korean or other consultants, they quickly chose the 38th parallel, a proposal which was then presented to and accepted by the Russians (Cumings 1997: 186f). The division of the country was thus haphazardly designed and implemented years before the actual separation was made official by the declaration of the two hostile republics in 1947. And with the United States and the Soviet Union becoming hostile rivals soon after World War II came to an end, both parties now started to engage in the massive financial, logistical, and ideological support of the (very few) fractions of politically involved Koreans that they saw as closest to their own interests.

The brutality of Japanese imperialism, in combination with the devastation caused by years of war, meant that Korea's nationalist movement lay in shambles in mid 1945. After liberation, no unified voice could be found among the surviving actors that could have countered the imposed division of the nation-state. The nationalist forces of Korea available at that time, as Kim Byung-Kook and Im Hyug-Baeg claim, "were individuals rather than organized political forces, isolated from society organizationally, separated from each other by personal ambition, and holding incompatible ideologies." Also, the fact that many of them were "geographically dispersed in faraway sanctuaries, and politically integrated into rival ideological blocs" meant that "the Korean nationalist movements failed to develop an umbrella organization capable of integrating different elite factions into a power bloc" (2001: 12).

Consequently, in the North, it would be Kim Il Sung who won the bid for power, a man whose rather poor (Presbyterian) family had gone to Manchuria in the 1920s in search of a better life, where the young Kim then joined the guerrilla movement and gained a reputation for his skills in battle (Lankov 2003: 49ff). In the South, it was Syngman Rhee, the son of an aristocratic family who had studied at both Harvard and Princeton before becoming a right-wing nationalist leader and *persona non grata* under Japanese colonial rule. Following the defeat of Japan, he returned from the United States where he had been in exile for 30 years (Kim B.K. and Im 2001: 12). Rhee's influence in the South, however, was stronger with the American provisional government than with the actual local population, it seems. Instead of flocking around Rhee and his Austrian wife, people had often already been drawn into the "Committee for the Preparation of Korean Independence," a political organization founded by left-wing nationalists that was initially very successful in mobilizing the population. Grassroots "Peoples' Committees" had been established all across the country after the Japanese departure. The interim American military government would spend most of its years in power seeking to suppress these committees and other left-wing initiatives. In this way, the ground was inadvertently prepared for continued turmoil and uprisings against the emerging leadership of Syngman Rhee (Cumings 1997: 192; see also Roehner 2014).

Just half a century after Korea had come to arrange itself with the downfall of one hegemonic regional power (China) and the rise of another (Japan), it saw itself ever more deeply drawn into conflicts that were no longer merely the regional ripples of global processes emanating from Western power centers. Instead, the new global powers—the USA and the USSR—would now send their troops directly to this country that was divided despite its undisputed status as a victim of Japanese imperialism. And while the last Soviet soldiers would leave the North in the 1950s, the US, it soon became clear, had come to stay, turning South Korea into a quasi-satellite state over the decades that was to form a crucial puzzle piece in the ever expanding "empire of bases" (Johnson 2004: 151) that it built up after World War II.

The Korean War and its Aftermath

On June 25, 1950, boosted by the communist revolution in China that had taken place less than a year earlier, North Korean troops crossed the 38th parallel and marched into the South. The United Nations (UN), under the leadership of the US, sent hundreds of thousands of troops to support the Southern side, while China and the USSR stepped in to support the North with over a million soldiers of their own (Chen 1994; see also Cumings 1981, 1990). This proxy war resulted in a stalemate, bringing the opponents back to the very same line that had divided the country before the military escalation started in 1950. Over the course of the conflict, however, large parts of Korea were utterly devastated. The overall destruction rate of North Korea's cities, for instance, is estimated to have been between 40 and 90 percent, with many areas completely obliterated in the course of the three-year-long carpet-bombing undertaken by the United States, which forced practically the entire population to retreat into tunnel systems in order to survive (Cumings 2004: 158ff). Up to 3 million civilians would perish throughout the war, and the provisional division of the country would be set in cement.

In the North, Kim Il Sung and his most trusted supporters (primarily recruited from the small number of surviving guerrilla fighters who had been with him in Manchuria) would now go about building a thoroughly anti-Japanese and anti-American, rapidly industrializing communist state. Over the decades, the North Korean elites added such a strong focus on military alertness into the mix that North Korea was gradually turned into arguably the most militarized state of the globe. Bruce Cumings has aptly described the North as "the world's most complete garrison state" (2004: 14)—a country espousing a kind of socialism of the barracks that would put any other socialist country with militaristic ambitions to shame with the sheer size, breadth, and depth of its military encroachment into civilian realms. In the South, the US had in the meantime also placed its bet on a strongman-in-the-making: Syngman Rhee. Prior to the outbreak of the Korean War, he had already ordered the quelling of an uprising on the island of Cheju that led to deaths numbering in the tens of thousands. And after the Korean War, oppositional politicians who could potentially pose a threat to Rhee's authority were also targeted, with the 1959 execution of Cho Bong-am, the leader of the Progressive Party who had won about a third of the vote in earlier presidential elections, figuring as a reminder to the population of just how authoritarian Syngman Rhee could be (Lankov 2011).

Through the distribution of foreign aid—ranging in the billions of dollars and, not surprisingly, mainly coming from the United States—Rhee had managed to buy himself a reasonably quiescent populace until Cho's death. To give an idea of the scale of the assistance the South received, it may suffice to note that in 1956, 58.3 percent of the country's total budget was made up of foreign aid, most of which came from the US. During that year, US economic aid peaked at $365 million (Savada and Shaw 1990), money that was primarily invested in the population's food supply, in technical assistance, and in the infrastructure that had been obliterated throughout the war (Lie 1998: 29). All the while, it was US military assistance that made up the most significant section of foreign aid: between 1953 and 1960, approximately $8.7 billion were spent on enlarging the security apparatus of the country (Congressional Budget Office 1997: 23).

Following the murder of opposition leader Cho, however, the notion rapidly spread among the population that the South's economic situation was not likely to improve any time soon under Rhee's leadership. Subsequently, the president saw his autocratic leadership increasingly challenged, leading to ever-growing student protests. The dramatic expansion of educational facilities undertaken by Rhee, which had been enabled by US donations (Brazinsky 2007: 189ff), now backfired in an impressive way. While in 1945, there had only been approximately 120,000 middle, high school, and university students enrolled in all of South Korea, their number rose to over 900,000 in 1960 (Adesnik and Kim 2008: 6), the year when Rhee was forced out of office and into exile. In spite of this victory of April 1960, the moment for a real democratic opening had not yet arrived: after a short-lived stint of democracy, a military coup would put a temporary end to South Korean citizens' efforts to gain political self-determination. General Park Chung-hee—for better or worse the most influential of South Korea's leaders in the 20th century (Kim H. 2005: 205)—would subsequently rule the country from 1961 until his assassination in 1979.

Figure 2.1   Military parade in Seoul on Armed Forces Day (October 1, 2008)

Militarized Modernity and Capitalism of the Barracks (1961–87)

> No special proof is necessary to show that military discipline is the ideal model for the modern capitalist factory. (Weber 1991: 261)

South Korea after 1961, molded according to Park Chung-hee's vision, was soon fully engulfed by a kind of "militarized modernity." This notion, introduced by Moon Seungsook (2005), seeks "to capture the peculiar combination of Foucauldian discipline and militarized violence that permeated Korean society in the process of building a modern nation in the context of the Cold War" (2005: 7). The contradictory and violent aim that the term "militarized modernity" points to was first endorsed by Park after his coup, achieved its most intense realizations in the 1970s, and slowly dissipated from the 1980s onwards (Moon 2005: 42). Under Park's iron fist, past humiliations at the hands of the colonial power (the Japanese had justified their interference in Korean affairs on the grounds of the supposed "backwardness" of the country after all) were now to be compensated by Korea becoming a strong player in the economic and military field. Park Chung-hee deliberately forged a developmentalist state by imitating past imperialist practices of exploitation, resulting in a form of capitalism of the barracks that entailed a fashioning of entire social sectors after military structures. In such a way, a more efficient domination of the potentially unruly people was made possible, while South Korea was propelled onto an unprecedented trajectory from the margins toward the center of the capitalist world system.

Park Chung-hee in many ways was the very embodiment of the militarized male figure in support of the nation that Sin Ch'ae-ho had dreamed of decades earlier, and he embodied many of the contradictions of this Japanese-imported phantasm, too. Born in 1917 in the Kyŏngsang region, he was the youngest of seven children of a poor farmer.[11] First trained as a primary school teacher, in his early 20s he had left Korea for Manchukuo to join the Imperial Army (Drennan 2005: 281), where he excelled to such an extent that he climbed the hierarchical ladder within the Japanese military system despite the "flaw" of his Korean birth: "A biography subsidized by his supporters noted how proud he was to get a gold watch from Emperor Hirohito as a reward for his services, which may have included tracking down Korean guerrillas who resisted the Japanese" (Cumings 1997: 355).

After independence, amidst the agitation created by the slow crushing of the people's committees by the US interim government, he had become involved in a rebellion and was arrested in 1946 on account of being a leftist; a somewhat ironic allegation when his future career was to make him one of the most dedicated anti-communists that Washington would ever have on its side. After the coup in 1961, which he carried out with only a handful of loyal colonels and 3,500 soldiers at his disposal (Cumings 1997: 352), the US authorities were momentarily worried about the change in leadership. Soon enough, however, they embraced the new strongman in Seoul—a person who proved to be exceptionally gifted in combining economic and military strategies introduced by both the former and the present-day powers-that-be in East Asia, and who shaped these tactics into something new that could benefit his particular vision of South Korea.

A Nation Being Built

The term "nation building"—a misnomer for state building that has again become a buzzword in the global political sphere since the early 1990s—is a rather ambiguous, but strangely accurate description of what was to come for South Korea under Park Chung-hee. Park was attempting to bolster the state *and* the nation, and this double approach proved to beget both great successes and new contradictions. Before his accession to power, South Korea was still a country very much weakened by war and division. Syngman Rhee had propagated a military solution to the country's division with his idea of a "march North," which he endorsed until his last days in office (Shin et al. 2013: 6), but Park put priority on consolidating the South Korean state instead, thereby postponing the question of a potential reunification to a later date when the more pressing economic questions would have been solved.

And economic issues were pressing indeed for the consolidation of his regime, especially as the omnipresent comparison with the North laid bare years of neglect in this particular domain. At that time, in contrast to South Korea, the North had quickly "overcome the wartime destruction and was steaming ahead of the South" (Cumings 1997: 352). Under Syngman Rhee, South Korea had been permanently mired in economic stagnation that was the inevitable result of widespread corruption in the country, a lack of foreign exchange reserves, and high unemployment rates to be found in all sectors of society. In the early 1960s, the country's per capita gross domestic product (GDP) stood at a mere $82, a strong indicator to the population that matters had to be changed rapidly (Adesnik and Kim 2008: 7). The yearning for reunification, which many believed would solve these economic problems, had become such a crucial political factor that, in the year before the military coup, large-scale student protests in support of immediate reunification had become the order of the day (Lee N. 2007: 28).

For Park Chung-hee, to stabilize his hold over power was a matter of survival, and, in addition to economic measures speedily undertaken, an appropriation of the nation (*minjok*) proved to be the most crucial strategy he endorsed:

> The term carried a great deal of legitimacy because of its association with Korean nationalistic movements of the late nineteenth century and independence movements under the Japanese colonial rule (1910–1945). Park's two main objectives were economic development and national security, and in order to achieve these goals, he appealed to the *minjok* sentiments of Koreans. As a result of the Korean War (1950–1953) and the strident anti-Communist propaganda during the post-1945 period, *minjok* gradually acquired an anti-Communist flavor. (Walhain 2007: 85)

The anti-communist nationalism propagated at this time was not only steeped in patriotic sentiments, but was also nurtured by widespread sentiments about Korea's putative backwardness compared to the West—a notion of underdevelopment that could only be cured by progressive Americanization in the eyes of many (Lee N. 2002: 47). In order to fight the supposed dangers of backwardness, economic competitiveness against the North was now made the number one goal, in this way both securing the support of the United States and cementing a further distinction from the brothers and sisters up North who were symbolically positioned ever further out of reach.

The Economic Push Forward

The 1960s consequently were marked by rapidly expanding economic activity. The first of four five-year-plans was drawn together hastily, and placed a heavy focus on industrialization that was to bring the desired increase in exports. At the beginning of these reforms, South Korean exports made up only 3 percent of the country's rather meager GDP; by the 1980s, this figure had increased to a staggering 33 percent (Krueger and Yoo 2002: 608, 609). All the while, the majority of Korean labor was still bound up in the rural sector, which meant that an urbanized labor force that could be employed in the newly built factories had to still be created through a strategy of systematic agricultural underdevelopment (Kim I.K. 2010: 114f; Lie 1998: 99). More urgently, however, the question of where to find the necessary financial resources to get the industrial sector jumpstarted in the first place needed to be solved. In this respect, Park depended on a double strategy of both local and foreign investment. After 1945, Syngman Rhee had sold the previously Japanese-owned banks in the country almost as quickly as they had fallen into the state's hands. Park, who clearly saw the benefits of keeping the financial sector within his reach, would now re-nationalize these banks, which allowed the state to dole out large sums of money to companies willing to invest in factories that promised to stimulate exports (Kim E. and Park 2011).

The companies that would profit the most from Park's generous credits handed out to auspicious industries were South Korea's *chaebŏl*—that is, family-owned and -managed industrial conglomerates that proved to be decisive for the rise of the country's economy. Modeled after the Japanese *keiretsu* conglomerates (many of which had been heavily involved in the grab for Korea's assets during colonialism), these companies were typically founded after World War II, and quickly rose to prominence in the years under Syngman Rhee (see, for instance, Chang 2006; Han 2008; Kang 1996; E. Kim and Park 2011; Kwon and O'Donnell 2001; Murillo and Sung 2013). Because Park Chung-hee was now able to buy *chaebŏl* compliance with governmental goals through his near-complete control of the financial sector, a strategic and highly successful alliance between state and industry was shaped that would lead the country on its path toward phenomenal economic success. While Korea was still one of the poorest countries on earth in the late 1950s, after the economic reforms in the early 1960s, "Korea began growing at sustained rates previously unheard of in world history" (Krueger and Yoo 2002: 606).[12]

All the while, foreign investment in Korea's emerging industrial sector played an equally important role. After his ascent to power, Park Chung-hee quickly turned his gaze in a direction that he was well acquainted with from his younger days. Being able to capitalize on his old ties with Japanese right-wingers, and with some further encouragement by the United States, Park soon initiated a hugely controversial Normalization Treaty with the former imperialist power, which secured an additional $800 million in developmental aid for South Korea. The subsequent protests led by students opposing the move, who were arguing that "the country's leadership had exchanged national pride for the promise of loans and economic aid from Japan" (Lee N. 2007: 30), were silenced by the imposition of martial law. This strategy of repeatedly declaring a national state of emergency whenever his rule was questioned by internal forces would become something of a trademark of Park during his 18-year-long rule (Gregg 1999; see also Cumings 1997: 358).

The other source of foreign investment that the regime would actively court was that of the United States. Billions of US dollars kept flowing into the country, just as they had under Rhee, and Park would continue what his predecessor had already managed quite adequately, too: that is, to milk the economic benefits of his country's subjugation to US security interests for all it was worth (Cumings 1997: 304ff). In addition to direct aid coming in from the United States, in 1965 another lucrative opportunity arose that resulted from the close alliance with the Americans. As the United States sent ground troops to Vietnam in the mid 1960s, South Korea, too, entered into the Vietnam War. Until 1973, more than 300,000 soldiers were sent to Vietnam to fight alongside the South Vietnamese troops, making the Korean contingent the largest body of foreign troops besides that of the United States.

"South Korea's submilitarism in Vietnam," writes Lee Jin-kyung, "was a significant factor in securing South Korea's position as a subimperial force within the U.S.-dominated global capitalism in the years following the end of the Vietnam War" (2009: 657). Indeed, the country's military engagement in South East Asia quickly started to pay off for Korea, with money from the Vietnam War becoming the largest source of foreign exchange earnings in the mid 1960s (Gregg 1999). Direct profit—totaling over $1 billion between 1965 and 1972—was also generated through remittances sent home by the Korean soldiers and by thousands of additional workers who had been taken to the war zone to fulfill the lucrative contracts that Korean companies had been awarded in South Vietnam (Lie 1998: 64). Hyundai, for instance, a *chaebŏl* founded in 1948, had first "managed to distinguish itself in the 1960s as a construction company by gaining US military contracts" (1998: 95), which, quite naturally, led to taking the next step into Vietnam (1998: 64, 95).[13]

With ever increasing US military activity in Asia, the American troops that had been stationed permanently in the South since the Korean War (after 1948, US soldiers left the peninsula for a short while, only to return in 1950) were now recognized as an ample source of the hard foreign currency that was so urgently needed to build up the economy. Katharine H.M. Moon maintains that US troops in the country may have contributed as much as 25 percent of South Korea's GDP in the 1960s (1997: 44). The large number of US soldiers deployed in Vietnam also meant many more rotational troops coming through the US military installations in South Korea, with many soldiers flying for Rest and Recreation (R&R) trips to Seoul only to return to the combat zones of Vietnam in a matter of days. This further contributed to the rapid expansion of red-light districts near US bases, where the dollar was the main currency that could buy the young soldiers any service imaginable.

Militarization of Daily Life under Park

In 1972, the South Korean constitution was amended, which marked the beginning of the *Yusin* era (1972–79). From then onward, matters for the South Korean population took a drastic turn for the worse: the declaration of martial law for the whole country that, hand in hand with the new constitution, resulted in closed universities for most of that year, with tanks placed at many locations throughout Seoul to intimidate the public (Chang 2008: 656). In addition to the usual crackdowns on potentially unruly dissident students and labor groups, young men who had grown their hair long in an attempt to emulate Western-type countercultures were now subjected to everyday police harassment (Lankov 2007: 326).

In the midst of a larger campaign to stop the putative corrosion of Korea's ethical and moral core that was supposedly undermined by all-pervasive Western influences (see chapter 5 for a longer discussion of this), the issue of masculinity once again moved to the center of some debates, with these young men now seen as a threat to the militarized masculinity that the leadership was embracing as an ideal.

Universal male conscription played a major role in this process of generating soldier citizens under Park Chung-hee. The universal draft had already been introduced under Syngman Rhee in 1948, but, during that earlier regime, evasion of military duties was an easy enough endeavor if one had money and connections. Conscription, therefore, was understood by many to be the "poor-men's-draft" rather than an egalitarian measure that affected all male citizens alike. Park Chung-hee would drastically change this state of affairs, and managed to bring down the evasion rate of 16 percent in 1958 to a negligible 0.1 percent in 1975 (Tikhonov 2009). Through draconian punishment doled out to those refusing to join in, the military came to be:

> praised as the way to make a "real man" (*chincha sanai*) on all levels of education, in mass culture, and in the media. Draft-evaders were made national scapegoats, accused of being both unpatriotic and unmanly, as manliness was now firmly identified with willing service in the army. This concentrated flow of militaristic propaganda, together with the strengthened popular legitimacy of the Pak Chŏnghǔi regime brought about by tangible economic success, seems to have won "ideological hegemony" of sorts for the conscription state. (Tikhonov 2009)

Farewell ceremonies for the soldiers shipped off to the battlefields in Vietnam—which resembled the public rituals performed during World War II under Japanese rule—were utilized to celebrate Korea's new-found military might. Park Chung-hee himself glorified his troops as capable of demonstrating "the bravery of Korean manhood to the world," and expressed that the country's inclusion into the Vietnam War efforts signified Korea's ascent as a "sovereign, mature adult nation" (Lee J. 2009: 659). While the opportunity to "remasculinize the patriarchal national community" (2009: 660) was fully exploited during such events, the physically fit soldier was now (once again) hailed as *the* leading figure for Korean-style progress. All the while, women and pacifists were relegated to the margins of this nation-in-arms in the making.

This martial vision of the South Korean (male) citizenry was increasingly turned into a reality through everyday practices targeting the population outside of the actual military barracks as well. Schools and universities made their students take part in military style exercises on a daily basis, a routine that Park had first experienced in Japanese-run Manchukuo in the 1930s (Han 2005). The working population, exploited in the country's ubiquitous sweatshops during nearly every waking hour, was subjected to morning drills and tough exercising, too; and their consumptive desires were curbed by the extraordinarily low wages they earned on the shop floor. Even when the economy eventually started to take off in the mid 1970s, the Park regime found ways to keep wages artificially low. The incorporation into the garment manufacturing industry of large numbers of young women who were subject to super-exploitation was one factor in this; another was that any attempts at unionization were criminalized and heavily punished (Lie 1998: 101).

With some of the longest working hours on the planet, workers' discipline was deliberately forged in *chaebŏl*-owned factories through imitating some of the practices and management ideas brought into the country by the US military (Lie 1998: 101). And universal male conscription, in addition to shaping a particular form of militarized masculinity, also became the primary tool that would facilitate the cultural revolution that turned Korean peasants into urban factory soldiers: "By imposing homogeneous training on South Koreans from every status and regional background, the military contributed to national integration and cultural homogeneity" (Lie 1998: 101), while also enabling the integration of large strata of society into the industrial sector (see also Janelli 1993). Curiously enough, it was precisely the sweatshops of Seoul that saw the initial moments of resistance to the nearly all-encompassing barracks that South Korea had become.

The First Few Cracks in the Barracks' Wall

Park's hold on power was inherently unstable and had to be buttressed by economic success. But ironically, this very success ultimately threatened his power. For Park's slogan "Let's Live well" (*chal sarabose*)—signifying in effect, "Let's live for once like the well-fed and well-clothed"— represented in essence the philosophy of a beggar, and people once out of beggary usually wish to live not by bread alone. (Paik 2005)

The near-total mobilization of South Korean society against the threat of communism penetrated practically every aspect of daily life during the 1970s. At the height of this repression, however, the beginnings of a new grassroots movement for change can be found. An act undertaken by a 22-year-old factory worker, Chŏn T'ae-il, proved to be the "single spark" (Cho 2003) that ignited a new movement for labor rights which inspired the oppressed working class. The young man was employed at a garment manufacturing unit located in Seoul's infamous Peace Market, a maze-like collection of several buildings that housed thousands of sweat shops, where predominantly young female workers labored under the most horrendous conditions.[14] Shocked by what he experienced at the Peace Market, Chŏn increasingly became invested in the promotion of labor rights—a dangerous task to undertake given the repercussions that trade union activists faced, with abductions, torture, and rape common practice in those days (Park M. 2005: 264). Growing desperate in his fight for the improvement of labor rights, on November 13, 1970, Chŏn took his own life at a protest near his workplace, in a much-publicized act of self-immolation. The incipient movement had produced its first major martyr—and, in Chŏn T'ae-il's name, the struggle for better labor conditions was continued, with the movement slowly but steadily gaining a committed following among the workers (Cumings 1997: 375f; see also Koo 2001: 69ff).

Not surprisingly, further crackdowns on those striving for more labor and civil rights were swiftly put in place. The rise of the labor movement coincided with major geo-political shifts that also put Park Chung-hee's regime to the test. While he saw his authority increasingly questioned internally, the Nixon Doctrine raised the pressure from outside, too. The Doctrine of 1969 stipulated that the US would increasingly leave the military defense of its major allies to their own armed forces, and a sharp reduction of US troops in South Korea followed. With Park's security arrangement significantly weakened in this way, he allowed talks to be held with the North on several occasions during the early 1970s. These resulted in a joint statement, issued in 1972, which outlined some commonly held positions on a future reunification of the two countries. However, once it became clear that South Vietnam was going to lose its battle against the communist North, Park Chung-hee and his regime walked away from these budding peace initiatives and focused on further entrenching their internal grip on power instead (Choi 2012).

It would take until the late 1970s for Park's rule to come to a sudden end. Following the violent crackdown on a hunger strike of 200 unemployed female workers in Seoul, protests broke out in Pusan, Masan, and several other cities, that were once again put down with another declaration of martial law. However, this time around, at the very height of repression, tensions within Park's regime would suddenly spike between the hardliners and those seeking a compromise with the protesters. Consequently, on October 26, 1979, Park would fall victim to his long-time friend Kim Jae-kyu, then director of the KCIA,[15] who shot him dead at a dinner party and later proclaimed at his trial: "I shot the heart of [the] Yusin Constitution like a beast. I did that for [the] democracy of this country. Nothing more nothing less" (*Korea Times* 2007).

De-Militarizing the Garrison State? (1980–)

The Minjung Movement

After Park Chung-hee's assassination, popular forces for democratization assumed that the moment of liberalization had finally arrived. The so-called minjung movement that had arisen since the 1970s recruited its followers mostly among politicized students, workers, and religious groups (Lee N. 2007). This movement, first and foremost, stood for a "particular postcolonial engagement with history" (Abelmann 1996: 20) that aimed to wrest the definitional power over the nation out of the hands of the generals again. Its actors, led by dissident thinkers and students, sought to "forge a broad class alliance among workers, peasants, poor urban dwellers, and progressive intellectuals against the authoritarian regime" (Koo 2001: 18). While in the 1960s and 1970s dissidents had sought reform as their main goal, the participants of this new 386-generation,[16] however, were radicalizing quickly amidst an atmosphere of harsh state suppression, with a revolution becoming the new, clearly stated goal of these predominantly young people (Park M. 2005: 266).

Influenced by both Marxist readings and the *Chuch'e* ideology of self-reliance in North Korea that they were exposing themselves to for the first time (Lee N. 2007: 109ff), minjung activists throughout the 1980s would focus "on organizing the downtrodden masses (minjung) in general and the working class in particular. Believing that the working class was the main historical agent of revolution, they strove to build links with industrial workers" (Park 2005: 75). This they sought to achieve through clandestine mobilizing actions in factories, villages, and shantytowns; in the early 1980s, for instance, several thousand university students would leave their campuses behind to take up jobs at factories, where they tried to engage with workers and establish unions, with the number of students secretly working at factories swelling up to an estimated 10,000 by the end of that decade (Park 2005: 275f). The left-wing students who opted to stay at the universities would often engage in street agitation, or organized themselves in clandestine study and reading groups with other minjung activists. Small-scale protests, usually involving 50 to a few hundred students, were held in different parts of Seoul on a regular basis, with the protesters blocking traffic, chanting slogans, and handing out leaflets to passers-by. In a "hit-and-run" fashion, they would quickly move to another location before the police had even arrived on the scene, thereby maximizing their impact while at the same time avoiding mass arrest (Park 2005: 280f).

In the turbulent atmosphere after Park's death, rallies were quickly expanding in size and gathering momentum across the country. With many of the student leaders imprisoned under Park Chung-hee now returning to the re-opened campuses (universities had been closed down for almost six months after Park's assassination), people took to the streets in masses, demanding an acceleration of democratic change. Finally, on March 15, 1980, 100,000 students were joined by another 300,000 ordinary citizens in central Seoul, all demanding that martial law be lifted immediately (Lee N. 2007: 45). In the meantime, however, a coup within the army, staged just two months after Park's assassination, was already threatening to bring yet another military strongman to the forefront. It had been organized by Chun Doo-hwan, a man who shared Park's peasant background, his origin from the Kyŏngsang province, and who had also gathered some first few military experiences in the Imperial Army in Manchukuo. Chun was quickly rising to the top among the military and intelligence elites who saw themselves entitled to power (Drennan 2005: 285f), while protesters continued to challenge such claims on a daily basis.

The Kwangju Popular Uprising and Speculations over US Complicity

It was at the periphery of the country that resistance to Chun's emerging regime turned out to be the strongest. In May 1980, protests in the south-western town of Kwangju (which is the home of long-time oppositional leader Kim Dae-jung) turned into a massive popular uprising within a matter of days. With up to 200,000 of the 700,000 citizens of the town taking to the streets, the seasoned paratroopers sent in to terrorize civilians were driven out by the population who, in the meantime, had taken up arms to defend themselves against the soldiers. Over the next ten days, this uprising would be crushed in the bloodiest way, leaving large numbers of people dead in Kwangju. The actual death toll is still subject to much debate: the regime spoke of less than 200 casualties, a number that was used by subsequent conservative governments all the way into the 1990s. However, Sallie Yea, among others, ascertains that the actual number may have been as high as 2,000:

> Officially, according to the Korean government, 190 people were killed in the uprising. However, unofficially, upwards of 2000 people were reported as dead. This figure is derived from Kwangju's monthly death statistics, which totaled 2600 for the month of May 1980, which is 2300 deaths more that the monthly average at the time. (2002: 1557)

The mass murder of Kwangju's protesters, an event that has been described as "Korea's Tiananmen nightmare in which students and young people were slaughtered on a scale the same or greater than that in 'People's' China in June 1989" (Cumings 1997: 338), forced the Chun government to go to the greatest lengths in order to stop subversive news of the uprising from spreading.

> The military burned an unknown number of corpses, dumped others into unmarked graves, and destroyed its own records. To prevent word of the uprising from being spoken publicly, thousands of people were arrested, and hundreds tortured as the military tried to suppress even a whisper of its murders. (Katsiaficas and Na 2006: 1)

Furthermore, for their putative violation of "public peace and order," up to 40,000 ordinary people were shipped to the infamous Samchŏng re-education facility during the years following the uprising, where (assumed) leftists were to be turned back into upright citizens (Cumings 1997: 384).

Despite such draconian measures, in the latter half of the 1980s, protests—which were typically led by students or workers—would continue in full force. And in the midst of this ever expanding social movement for democracy, labor, and civic rights, slowly "the words 'Yankee, go home' became stock phrases in the dissident arsenal" (Cumings 1997: 338). Indeed, US support of the Chun Doo-hwan dictatorship that, in yet another military-led bid for power, had sent troops against its own people, proved to be the turning point for participants in the minjung movement, who would now openly, and at times violently, express their anger over the United States' entanglement with the regime. The main accusation against the American ally now became that of complicity in the Kwangju murders. As the supreme command over the Korean military had been in the hands of a US general at that time, protesters argued, Chun could have never initiated the large troop movements against Kwangju's citizens without US authorization.[17]

The very ferocity with which such anti-American sentiments washed through South Korea now—a place that, after all, was well known for its seemingly widespread public support of the United States—took many analysts by surprise. Lee Namhee argues that these heightened sentiments can be explained by the widespread anticipation that, after Park's death, the United States would come down on the side of the people: "The public's historical image of the United States as an ally of the Korean people was such that the people of Kwangju expected the United States to intervene on their behalf during the uprising" (Lee N. 2007: 51). However, at the very height of the Kwangju uprising (during a National Security Council meeting on May 22, 1980), the US leadership opted for an approach instead that entailed "in the short term support [of the Chun regime], in the longer term pressure for political evolution" (Adesnik and Kim 2008: 18). Less than a year after Kwangju, the newly instated Reagan administration even chose to cordially host Chun in the White House,[18] and many Koreans felt a deep sense of betrayal during such blatant public displays of US support for their new dictator. In a way, though, the US was made to take the blame by the Korean generals just as well, who saw an easy opportunity to wash their hands, at least partially, of the blood they had spilled of their own citizens (Drennan 2005). Regardless of the putative or actual involvement of the US military in the quelling of the Kwangju uprising, in this decisive moment the United States proved to be unwilling to intervene publicly on the side of the actors for democratic change, who were for now on their own in their struggle.

Democratization and the Thawing of North–South Relations

In June 1987, a point in time was reached when the minjung movement succeeded in mobilizing enough sectors of society outside of their core milieu of battle-seasoned campus warriors. Earlier that year, a 23-year-old university student had died from the consequences of police torture while imprisoned for his participation in the movement for democratization. His death would only become known months later, in the midst of protracted discussions over Chun's announcement that he would directly hand his presidency over to his second-in-command, Roh Tae-woo. Taken together, these two incidents caused so much outrage that people from all walks of life now took to the streets in order to join the mass protests that went on for several weeks. With millions of people marching in what came to be known the "June Democracy Movement," dictator Chun was eventually forced into giving democratic concessions and allowing general elections for the role of president to take place. However, due to an unfortunate split in the opposition, which divided the dissidents into two antagonistic camps, Roh Tae-woo—the next and last military man in line to walk into South Korea's presidential office—won the elections in late 1987 and ruled the country until 1993 (Adesnik and Kim 2008).

Only with the subsequent election of Kim Young-sam would the first non-military ruler in 32 years take charge of South Korea's presidency. Kim, during his time as head of the state, forced both Chun and Roh to stand trial for their past actions during the Kwangju uprising. Chun, convicted of treason, mutiny, and other crimes, would receive a death sentence, while Roh was sentenced to many years in prison; both men were pardoned one year later by incoming president Kim Dae-jung. After Kim Dae-jung's successful bid for the presidential office, the first non-violent transfer of power from the ruling to the opposition party in South Korean history took place, and marked the consolidation of the Republic of Korea's democracy. Among the most central elements of Kim Dae-jung's presidency[19] would certainly be his Sunshine Policy, through which he sought to find a new way to approach the Democratic People's Republic of Korea.

At that time, the vicious struggle for economic supremacy between the two halves of the formerly united country had already been settled firmly in favor of the South, a fact that even the Asian currency crisis of 1997, which hit South Korea quite hard, could not reverse. After the collapse of the Soviet Union and the Eastern bloc, North Korea's most important trading partners simply vanished into thin air and its economy consequently shrank at a dramatic rate. The heavy industries of the country needed to be fueled by oil, electricity, and other imports that were no longer arriving. A series of natural disasters worsened the situation for the agricultural sector, and rampant economic mismanagement, coupled with enormous expenditures that went into sustaining the large military apparatus, further strained the country's economy (Cumings 2004). With its population in desperate need of food and other supplies, the unexpected death of the country's leader Kim Il Sung in 1994 further heightened a sense of crisis and led to ever greater economic, military, and political dependency on China under the new leadership of Kim Jong Il.

Faced with such turbulence and the subsequent famine that ravished many parts of North Korea in the 1990s, both Kim Dae-jung and his presidential successor, Roh Moo-hyun (2002–7), opted to send food donations and other developmental aid up North. While aid became an integral part of the Sunshine Policy, which sought finally to resolve the conflict with the North, the message was sent across the globe that peace was perhaps possible between the contestants now that one of them seemed to be on the brink of collapse. The widely publicized meetings of Kim Jong Il with Kim Dae-jung and Roh Moo-hyun, which took place in 2000 and 2007 respectively, functioned as symbolic cornerstones of these efforts toward reconciliation, which were augmented by significant joint economic projects such as the Kŭmgangsang Tourist Region,[20] the Kaesŏng Industrial Zone,[21] or the re-building of a railroad to connect the two Koreas.[22]

While US President Bill Clinton, throughout his two terms as president, was an adamant supporter of renewed relations between North and South Korea and of a de-escalation of the old conflict, when George W. Bush (2001–9) marched into the highest office of the United States he embraced a decidedly anti-North Korean agenda. This would culminate in Bush's inclusion of the country in his infamous "axis of evil," a term he coined during his State of the Union Address in 2002. The increasingly paranoid military regime in North Korea used such saber-rattling by the US to speed up the development of its Nuclear Weapons Program, leading to a first nuclear test on October 16, 2006 (followed by additional tests in 2009 and 2013). Given how Bush actively undermined the peace efforts made through the Sunshine Policy, it is not surprising that relations between the US and South Korea deteriorated at that time as well. A new low point was reached between the two countries under the leadership of former human rights lawyer Roh Moo-hyun, who was elected as president in 2002 after a decidedly anti-American campaign, during which he vowed, among other things, "to create greater distance between Washington and Seoul" (Feinerman 2005: 215).

In addition to the larger geo-political shifts triggered by the "Global War on Terror," the day-to-day agitation of a dense network of left-wing South Korean non-governmental organizations (NGOs)[23] and the more event-based "citizens' movements" (Shin K.Y. 2006) proved to be crucial factors in the re-imagining of the role of the United States that was occurring in South Korea. Many of these civic groups directly grew out of the minjung movement, which had slowly transformed itself into a much more diverse grid of social movements since the late 1980s (Abelmann 1997; Kim D.C. 2006). The United States, it was now argued by some of these organizations, had been decisive in bringing about the Korean War with its choice to divide Korea in the first place. Through their willful partition of Korea in 1945, so the argument went, the US had set off a whole series of events, ushering in "anticommunism as the [South Korean] state ideology and extinguishing the once-vibrant post-1945 social movements" (Lee N. 2007: 42). In light of how strongly the United States and the anti-democratic regimes of Rhee, Park, and Chun were already connected in people's views, it literally only needed one event in the chaotic early 1990s to bring all these stored-up resentments to a boil. And when, in 1992, a Korean sex worker was murdered in a savage way by a US soldier, such a moment had arrived.

Anti-Americanism amidst Democratic Change

During the slow transformation from dictatorship to democracy in the 1990s and 2000s, the continuous United States Armed Forces presence in South Korea came under much fire, as it putatively attested to the unequal power relations between Korea and the United States. The murder of a young Korean prostitute who had lived and worked in Tongduch'ŏn, and who was brutally killed by an American soldier in 1992, would cause a particularly intense outbreak of negative sentiments that were hurled at US troops all over the country. During months of political agitation that followed this violent incident, the dead woman in question, Yun Kŭm-i, emerged as a central symbol in nationalist narratives for the putative suffering of Korea, the nation. Such an appraisal occurred precisely at a time when the systematic rape of Korean women by Japanese soldiers during World War II was also becoming widely known. It was easy for anti-US military activists at that time to point out the similarities between the so-called comfort women system, constructed for the benefit of the Japanese Imperial Army, and the contemporary red-light districts in South Korea that US soldiers frequented in their free time (Moon 1999). To be sure, the Yun Kŭm-i murder—which will be discussed in greater detail in the following chapter—was not the first violent event involving US soldiers and Korean citizens, and it would also not be the last. But it was indeed the one occurrence that first triggered protracted public turmoil; and a pattern was established back then that would be followed whenever a new violent transgression involving US military personnel or one of their dependents came to light.

In April 1997, for instance, two young US citizens—Arthur Patterson, who was the son of a US serviceman, and Korean-American Edward Lee—stabbed to death a Korean college student, Cho Jung-pil, whom they happened to randomly come upon in the public restroom of the It'aewŏn Burger King restaurant. This murder case, too, received quite a lot of attention: the two suspects, subsequently blaming each other throughout the investigation and trial, both went free, and the commotion over the murder trial turned the It'aewŏn neighborhood into a no-go zone for many Koreans for years to come. And ten years after the Tongduch'ŏn killing of Yun Kŭm-i, a tragic accident in Yangju, a predominantly rural area north of the capital, left two Korean middle school students dead. Sim Mi-sŏn and Sin Hyo-sun, both 13 years old, were on their way to a birthday party when they were run over and instantly killed by a US military vehicle that was taking part in a training maneuver. When the two drivers involved—Sergeant Fernando Nino and Sergeant Mark Walker—were cleared of negligent homicide charges in a US military court in November of 2002, protests against the US military quickly spread once again, bringing thousands of middle and high school students to the city center to hold candle-light vigils for the two dead girls, with other concerned citizens joining in over the following weeks.

As in the case of the Yun Kŭm-i murder earlier, the publication of several pictures, depicting the badly mutilated corpses, played a major role in further enraging the public against the US military, with a leftist nationalist NGO apparently leaking the photos to further instigate uproar.[24] These protests peaked on December 14, 2002, with an estimated 300,000 people taking part in demonstrations across the country. The conservative newspaper *JoongAng Daily* reported that day that "45,000 people gathered in front of Seoul's City Hall to hang 'Yankee go home' banners, chant slogans like, 'Revise the SOFA,' 'Bush apologize' and 'Bring Mi-sun and Hyo-son back alive.'" In addition, the protesters "also sang obscenity-laced anti-American songs and tore several huge U.S. flags to bits before unfurling a Korean flag to shouts of 'We will recover our national pride'" (Min 2002).

An anti-American element was certainly also a contributing factor to the beef protests against the Lee Myung-bak government that took place in spring 2008. These protests were in fact the largest such demonstrations since the beginning of South Korea's democratization 20 years earlier. They were triggered by a controversy over an issue that many Koreans viewed as yet another infringement on their nation's fragile sovereignty: the negotiation of a beef deal within the larger framework of the Free Trade Agreement between the US and South Korea, and the subsequent reopening of the Korean market to US beef imports, which were generally viewed as unsafe by the public. The storm that broke loose after the details of the rapidly concluded negotiations became known has to be understood as a reaction to the general oppressive climate of the "latest chronic Korea disease, a fusion of authoritarianism and neoliberalism" under the leadership of conservative President Lee Myung-bak (2008–13), Pak Noja (Vladimir Tikhonov) argued at that time. Furthermore, he explained that these protests were an outcome of decades of quasi-imperial exploitation by the United States: "[T]his is a country under the military protection of the American empire, one that has had its pride hurt numerous times before, [with] the humiliating deal hurt[ing] the last remaining pride of its citizens" (Park N.J. 2008).

Despite the fundamental democratic successes of the 1990s and 2000s, under President Lee Myung-bak the ongoing conflict with the North once again took a turn for the worse. And, indeed, conservative President Park Guen-hye (the only child of military dictator Park Chung-hee), who entered the presidential office in 2013, has also followed Lee's hardline policies toward the North. With most of the concessions made to North Korea under Kim Dae-jung and Roh Moo-hyun by now reversed, the slow process of de-militarization of South Korean society that went hand in hand with these earlier peace efforts seems to be in an ever more fragile state, too. For instance, since Lee Myung-bak's presidency, riot police, who are in fact predominantly made up of young men doing their mandatory military service, have been dispatched in huge numbers to any putatively dangerous public event. From small human rights festivals to commemoration services for a former president,[25] their black combat uniforms could be seen scattered across the inner-city space on practically any day of the week after the febrile spring and summer of 2008, while I was undertaking my field research in Seoul.

Furthermore, under President Roh, the reduction of the mandatory time for male conscripts had been proposed—another plan that was retracted under Lee in an apparent bid to strengthen the country's military power once again. As a consequence, the introduction of an alternative service for those unwilling to take up arms in the name of the nation has also been postponed indefinitely. Thus South Korea today still holds the sad record of having the highest number of conscientious objectors imprisoned

Figure 2.2   Performance against mandatory military service in South Korea undertaken by pacifists in downtown Seoul

worldwide, a fate that over 15,000 young men have already experienced in South Korea since 1939 (Tikhonov 2009).

Fortifying against China's Rise: Strategic Flexibility and US Troop Realignments

"The United States," international relations analysts often like to argue today, "is increasingly concerned about the rise of China and its adverse effects on American's regional posture in Asia" (Chung 2006: 3). China's rise as an economic super-power has proven to be a real game-changer for relations in North East Asia. The times now seem long gone when the United States was still South Korea's main trading partner; in 2003, China first surpassed the US and has since firmly taken hold of the position of largest trading partner (Kalinowski and Cho 2012: 247). In 1992, the total trade volume between the two countries still stood at a meager $6 billion. By 2013, however, this number had changed to $270 billion, with Chinese–South Korean trade now worth more than the value of Korea–Japan and Korea–US trade combined (Jin 2014). A free trade agreement signed in June 2015 between the Republic of Korea and China is set to further cement these tight business relations over the coming years.

This expansive economic cooperation is a key facet in an ongoing geo-political shift with unpredictable ramifications for security arrangements both on the Korean peninsula and the Asia-Pacific region as a whole. The recent return of China as an economic super-power, after all, is accompanied by Chinese efforts to boost its military strength in such a way that it may match up to its newly found financial strength one day. In light of China's skyrocketing spending on its defense budget, the old hegemon of the USA increasingly feels the need to further entrench its military power in the region. Consequently, the realignment of US Armed Forces in North East and South East Asia has turned into a vast and ongoing project. President Barack Obama's much discussed "pivot to Asia" has already had a significant impact on the island of Guam, which is increasingly becoming the center of US military gravity in the Asia-Pacific region (Lutz 2010). The Philippines, too, which also finds itself threatened by China's military and territorial ambitions, is nowadays seeking to strengthen its military partnership with the United States— after several decades of relative independence from its former colonizer (Simbulan 2009).

South Korea, too, is increasingly drawn into this large-scale scramble for influence in Asia that has pitted the United States against China. This can be directly ascertained through the fluctuating numbers of US troops stationed in Korea. During the Bush presidency, the 37,000 troops previously stationed in the country were drastically reduced, eventually bringing the number of soldiers on the ground down to 28,500 (Feinerman 2005: 216). During the same era, the US forces in Korea announced their intention to "reduce its footprint" in urban areas[26] and to relocate all of its troops and facilities from the northern parts of the country to the areas south of Seoul. Such a move would have eventually resulted in the abandonment of all US installations near the cities of Tongduch'ŏn and Ŭijŏngbu, while most of the US facilities near P'aju—the city closest to the DMZ—have already been given up over the last decade. The 2.5 square kilometers of valuable land in central Seoul, on which the US Army garrison Yongsan (which also functions as the main headquarters of US Forces Korea) is located, was also slated to be handed back to the Korean government.

As of 2015, however, much of the relocation process still exists only on paper, while the actual expansion of facilities nearby P'yŏngt'aek (that are supposed to absorb the US troops to be relocated from the northern part of the country) has been steaming ahead, leading to a de facto increase in US military activity in South Korea. Furthermore, a distinct increase in rotational US troops who are temporarily making their way into Korea has also been reported (Rowland 2014)—another indication that American military involvement on the Korean peninsula may actually be on the rise rather than slowly abating. In the midst of such developments, China's close proximity to South Korean shores seems to slowly take a top position on the list of US concerns pertaining to the Korean peninsula. If geographical choices on where to erect military bases are any indication, the question of how to deter China's rise seems to have overridden the old worries over the supposedly vast threat that North Korea poses. The larger P'yŏngt'aek/Osan region (located about an hour and a half away from central Seoul), which is to hold most of the US troops scheduled for relocation, is after all a strategic location which brings China, and not North Korea, within closer reach of US weaponry.

These developments, taken together, give a strong indication that military deterrence still (and yet again) seems to be the key strategy embraced by most actors who have stakes in the Korean peninsula. In a sense, one old truism seems to hold some legitimacy when applied to Korea: the more things change, the more they stay the same. The Korean peninsula, with all its unsolved legacies of armed conflict, its continuous military build-up, and its geopolitically distinct location (perpetually situated in the midst of too many cross-hairs), will remain a heavily contested territory for many more years to come, it seems; a dynamic zone in the heart of North East Asia where regional and global powers, old and new, continue to make their bids for influence.

3

"The Colonized Bodies of Our Women ..."

Camptown Spaces as Vital Zones of the National Imagination

"Our People United!"

On October 28, 2008, a group of about 30 protesters, the majority of whom are middle-aged men wearing suits appropriate for a commemoration ceremony, assemble near the United States embassy in central Seoul. They carry a large banner with them that has the name of a woman written on it: Yun Kŭm-i. She was a 26-year-old South Korean sex worker who lived and died in Tongduch'ŏn, a town that is infamous across the country for its many US military installations. The woman's death 16 years earlier at the hands of a US soldier is remembered on this October day in manifold ways: several speeches are given, unification songs are sung, chants and slogans repeated with fists raised in the air. One peak of emotion is reached when a group of younger protesters brings a poster to the forefront that shows the Korean peninsula in its divided form, with the entire Southern part covered by stars and stripes and the words "US Army" written across the territory. In a common effort, the students tear the poster to shreds and bring into visibility what lies underneath: a united Korea that is equipped with arms and feet which it uses to kick the Americans toward Japan. "Our People United!" is written in Korean above the image of the peninsula, "Withdrawal of US soldiers!" can be seen below.

In nationalist narratives across the globe, the boundary of the nation is frequently imagined to be equivalent to the actual female bodies of said nation. In the South Korean case to be explored here, one woman's (dead) body came to play a crucial role in re-imagining relations between the Korean nation and its long-term ally, the United States of America. Violent acts of US soldiers stationed in South Korea have, since the death

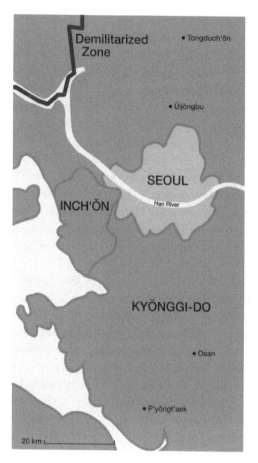

Figure 3.1 Map of the Greater Seoul area

of Ms. Yun in 1992, repeatedly been amplified by an outraged Korean public as symbols of the unequal relationship between the United States and this small East Asian nation. Crucial controversies in the 2000s, as we have seen, included that over the deaths of two teenaged middle-school students, Shim Mi-sōn and Shin Hyo-sun, who were killed by a US military vehicle in the summer of 2002, as well as the lengthy row over the destruction of the village of Taechuri (Yeo, 2006, 2010) to make space for the expansion of the US military Camp Humphrey, which will be further discussed in chapter 6.

By 2007, when I arrived in the country to conduct ethnographic field research on the US military presence in South Korea, the biggest protests against US troops in the country—which, as we have seen, had at times brought several hundred thousand people into the streets—had already become history. But during the nearly two years of research that followed, I was continually surprised by how many of my conversation partners would still depict GIs as potentially aggressive (sex) offenders at large in their city's entertainment districts. A 23-year-old male Korean student I interviewed in 2009, for instance, explained to me that whenever he goes to It'aewŏn, a district in Seoul that is located next to a US military base (see chapter 5), he experiences great anxiety because of the presence of foreign soldiers there: "I think this is really a psychological thing that I went through when I was still really young," he tells me, "because of all these US military crime reports. Because there are lots of GIs in It'aewŏn, and when I walk around the small streets of It'aewŏn, even during bright day time, I'm in total fear." He continues: "Just to imagine that there is this GI that beats me up and rapes my girlfriend, and the feeling that I wouldn't be able to do anything about it if that happened ..."

Statements like these gave me a sense of how deeply people's perceptions of US soldiers had taken a turn for the negative in this country that was once known as one of the most US-friendly nations in the world (Kim J. 2001: 173), and how much issues of gender were entangled in this process. Often directly referring to news stories about violent events involving GIs and Korean women, my conversation partners' narratives on the putatively "ordinary" behavior of US soldiers in Korea typically involved three components: an aggressive actor (that is, the foreign soldier, who was always imagined to be young and male), a victim (typically a young Korean woman), and an entertainment space of ill repute. Seeking to explore the emergence and the meaning of this seemingly set-in-stone discursive trinity of violent agent, female victim, and the "tainted" terrain in which offenses occur, I had to go back to the first critical event—the Yun murder that precipitated this particular view of US soldiers. An inquiry into this murder and its aftermath, I believe, allows me to tackle one question in particular: How and why did it happen that the figure of one murdered woman was turned into such a powerful stand-in for a nation perceived to be under constant duress? And which processes were at work in this particular "transformation of individual biography into social text" (Das 1997: 10) that this murder case would give rise to, allowing the emergence of a new, highly successful nationalist frame on the US military presence in the country?

In what is to follow, first the details of the murder and its ramifications will be laid out, with some attention paid to the nationalist discourses applied to the event. Second, I will explore the social and economic factors that have shaped the particular entertainment space in which the killing took place, turning it, over time, into an area that is both endangering and endangered on the margins of South Korea. In a third section, attention will be paid to the particular role camptown areas have played in the imagination of some writers of the minjung democratization movement. In the fourth section, anthropologist Marshall Sahlins' notion of structural amplification will be applied to explain how these marginalized spaces near US bases came to be understood as the very locations where US domination is realized. Through the extension of one seemingly private occurrence into an event of countrywide interest, we will see, specific places (that is, the entertainment areas frequented by US military personnel) were turned into imagined spaces of both national shame and transnational realms of empire by actors of the South Korean left. The sex workers employed in these spaces, as I will show through a vignette from kijich'on in the fifth section, have been stigmatized because of their miscegenation with US servicemen, a blemish that is also passed on to their (mixed-race) children. Finally, we will return to how these same women were formed into symbols of a ravaged nation, whose actual experiences were silenced in favor of a streamlined nationalist narrative that proved easy to think and act with in the midst of a perceived crisis over sovereignty.

"Our Nation's Daughter"—the Yun Kŭm-i Murder

During the last century the dream of money to be made in the company of American soldiers—and the possibility of marrying one of them and leaving South Korea behind for the United States—was the motor for female migration from the Korean sweatshops to the US military camptowns. Yun Kŭm-i was one of these young women who left a strenuous job at a factory for a ticket to Tongduch'ŏn, a town adjacent to a US base where she was looking for employment and opportunities in the GI clubs of the area. On the night leading up to October 28, 1992, she would get into a fight with Private Kenneth Markle, a then 20-year-old medic from West Virginia, after they had encountered another soldier she had been with the night before. Enraged, Markle bludgeoned her face and body with a coke bottle and sexually assaulted her with several objects. After she passed away from heavy bleeding, he threw detergent over her body in an apparent attempt to cover up his traces. Later in the morning, Yun Kŭm-i's mutilated corpse was found by her landlady. It would take two more days until Markle was arrested at the same club where he had been seen with Ms. Yun during the night of her murder.

Due to the gravity of allegations against the American soldier, the US Armed Forces, unlike on previous occasions, did not contest local legal authority over Markle, who would become the first US serviceman to be put on trial in a South Korean courtroom (Cho 2008: 7).[1] The most significant reason for this restraint displayed by the military, Kim Hyun S. explains, was the mounting public pressure in Korea:

> Previously, out of the estimated 40,000 crimes committed by US soldiers against Koreans, only 200 [suspects] had been handed over to the Korean court system. The murder of Ms. Yoon resulted in the first time that a US soldier was prosecuted in the Korean courts, and this was due solely to the massive protests that erupted in response to the horrific details of her death. (2009 [1997])

A critical factor contributing to the massive outrage over the murder was indeed the leakage of one picture to the press, depicting with full force the brutality of the killing. The gut-wrenching photo, apparently taken by crime investigation staff, shows the naked, bloody corpse in full exposure, legs spread wide open, with an umbrella still inserted into the victim's rectum. Rapidly disseminated across the country via news reports, this graphic image of the murder was later displayed at protests in front of US bases, and even served as a basis for discussion in some classrooms, as one Korean acquaintance born in 1982 told me, whose teacher relayed all the ghastly details of the killing to his shocked 10-year-old students, telling them at the end of his account of the murder to stay clear of US military areas.

The violent re-imagining of both actors and spaces affiliated with the US Armed Forces that was rapidly initiated after the murder heavily depended on the reproducibility of this one image, and on the photo's potential to travel across the space of the nation. Anthropologist Arjun Appadurai has alerted us to how images play into the imagination, arguing that one cannot fully be grasped without the other. "The world we live in today," he writes:

> is characterized by a new role for the imagination in social life. To grasp this new role, we need to bring together the old idea of images, especially mechanically produced images (in the Frankfurt School sense); the idea of the imagined community (in Anderson's sense); and the French idea

of the imaginary (*imaginaire*) as a constructed landscape of collective aspirations, which is no more and no less real than the collective representations of Emile Durkheim, now mediated through the complex prism of modern media. (1996: 49)[2]

In the pre-internet era of the early 1990s, Korean activists still had to depend upon regular print-media to disseminate the harrowing picture of the murdered woman, but even under these circumstances, the image was circulated widely across national territory. And, with outrage over the details of the killing mounting, news of this murder case would not quietly dissipate like the many times before. Immediate political mobilization was undertaken by non-governmental organizations (NGOs) of the left-wing spectrum (Feinerman 2005: 207). In the previous chapter, I have explored how a dense network of such groups had begun to emerge on the political scene ever since military dictator Chun Doo-hwan had been pressured into stepping down to make room for the first democratic elections in 1988. Turebang, a small group of Christian feminist activists (that I also volunteered for in 2009) was already working in the camptown areas at that time. In fact, this group had been around even before democratization: founded in 1986, the activists of this organization, with close Presbyterian ties, had built up several counseling centers for women working in the entertainment industry near US bases, including one that they ran in Tongduch'ŏn at the time of Ms. Yun's death. When Yun was murdered in the same town they worked in, the group, in cooperation with other women and religious organizations they were affiliated with, organized a protest against US military-related violence against women. With 3,000 people attending this first rally, much to the surprise of the groups that made the initial calls, the movement quickly spread throughout South Korea.[3]

The grounds for such widespread contestations had also been prepared by another closely linked subject that had recently received much publicity as well: just a year earlier, the so-called "comfort women" issue had reached a pivotal moment, when Kim Hak-soon, a South Korean woman then 76 years old, had come forward in the summer of 1991 to give public testimony about her former life as a sex slave for the Japanese Imperial Army during World War II. Sold into prostitution at the age of 16 by her stepfather, who had sent her to Manchuria, she had to serve a platoon of Japanese soldiers for the duration of five months before she could flee from the "comfort station" she had been assigned to. Encouraged by Mrs. Kim's example, hundreds of women from Korea and elsewhere were now following suit, coming forward to speak about their experiences of sexual exploitation as well, with their stories still making headlines in South Korea in 1992. It was indeed not hard for activists, then, to highlight the parallels between the "comfort women" system built up for the Japanese military, and the local sex industry that was currently servicing US Armed Forces personnel stationed in the country.[4]

Due to these factors and circumstances, tens of thousands of people would now follow the multiplying rallying calls of religious, feminist, and nationalist organizations to take to the streets in order to demand that Kenneth Markle be punished severely by a Korean court, and that the Status of Forces Agreement between Korea and the US be revised. This bilateral executive agreement, first signed in 1967 and revised in 1991 and 2001, established the legal framework under which US military personnel operate in Korea, thus clarifying to what degree the domestic laws in the country are to be applied to US soldiers while on Korean soil. Before significant amendments took place, one continuous source of local anger was the fact that, in the case of even the gravest violations, such as the rape or murder of a Korean individual, an apprehended US soldier would usually automatically be handed over to the US military authorities, who could then single-handedly decide on whether and how to prosecute the offender (Feinerman 2005; Mason 2009).[5] Within this context, after Yun Kŭm-i's death:

> [s]tudent groups staged violent protests while businesses, such as Korean restaurants and taxis, boycotted members of the US military. Some activists founded the "Joint Commission for Countermeasures" in order to investigate the murder case publicly. They organized press conferences, visited US military bases and demanded a public apology from US authorities. (Kern 2005: 261)[6]

In the heated atmosphere surrounding the trial, Markle was eventually sentenced to life imprisonment, with the sentence quickly being reduced to 15 years, which again sparked some accusations of undue US interventionism on behalf of its citizen (Kim N. 2008: 73).

Ostracized by Korean society as a prostitute while still alive, Yun Kŭm-i was posthumously tightly embraced as a daughter of the Korean nation during the months following her death and the trial of her murderer. The writings of the Committee on the Murder of Yun Kŭm-i by American Military in Korea are paradigmatic of the kind of incorporation of her death into a nationalist narrative that was quickly coming to override any other interpretation of the story of her demise: "Yun's mutilated body was material evidence of imperialist violence against the bodies of Korean women. These bodies were allegorized as the 'victimized' and 'suffering' Korean nation.... The body of Yun Kǔm-i became a metaphorical boundary for the nation" (Kim H. 1998: 189). She was described as "the daughter of [a] poor family," "our (the Korean) daughter," "a female factory worker," "poor prostitute," and "our nation's daughter who dreamed of America"; her life was remembered as emblematic of those of many other Korean prostitutes: "Under the Stars and Stripes, the colonized bodies of our women are thrown about," "how did you get here, Kǔm-i?" (Kim H.S. 1998: 190). In the imagination of pastor Chŏn Usŏp, Yun Kǔm'i even became a symbol for the slow demise of an entire nation: "The death of Yun Kǔm-i is not the death of an individual. It is the death of national sovereignty; the death of national (human) capital." Revitalization of the dying Korean body, according to him and others, could only be achieved by driving out the American troops (Kim H. 1998: 191). Before exploring camptown fiction—one particular predecessor of these nationalist appro-priations of camptown violence—in greater detail, I will now look at the particular history of the place in which the murder occurred: the military camptown area of Tongduch'ŏn.

Tongduch'ŏn as Endangering and Endangered Space

When night falls in camptown, the boys start to emerge from the gate. Some guys come all by themselves, their eyes hungry for some action. Others walk in groups of two, three, four, they wander down the main street, laughing, pointing, gazing. Some hold the hands of their [...] wives, with the occasional kid in tow, that's how they enter the camptown that clings to the walls of their military base. Blacks, Whites, Latinos, the occasional Asian American—the one thing that makes them immediately discernible as GIs is their trimmed hair. Many of them tattooed in more than just one place, most of them in good shape, tall, muscled, healthy, they radiate confidence and aggressive energy. They sit down at a bar for a beer and some food, chat with each other or talk with the former Korean sex worker that runs the place. Other boys head further down the street, straight to one of the clubs, to target the young Philippine woman of their choice that is waiting at the bar, all made up in a skimpy little dress. Slowly, the town awakens from its badly needed sleep, because daytime is only used to nurse the hangover, to repair the damage, to fix what can be fixed; night time, again and again and again, to get wasted, ruined, laid. (Field diary entry, June 2009)

The town of Tongduch'ŏn, in which the Yun murder took place, is mostly characterized by its deep and long-term geographical, social and economic exclusion from the rest of South Korea. Tongduch'ŏn, though located at the very heart of the Korean peninsula (approximately 30 km north of Seoul's city boundaries), was, after the division of the country into two antagonistic halves, relegated to the status of a border zone near the dead space of the Demilitarized Zone (DMZ). Prior to the division of 1945, the area had undergone some modernization because Japanese colonialists had used it to feed their large-scale military enterprise. In this way, a small town had emerged at the beginning of the 20th century close to the roads and railways built in this region that were to enable the exploitation of natural resources that made a contribution to fuel Japanese war efforts. This existing infrastructure was possibly the key factor as to why the US Armed Forces chose this particular location later in order to build up a net of military bases after the end of the Korean War (1950–53) as well (Kim B. 2007: 21).

After active combat between the Northern and Southern forces and their respective allies had come to an end with the signing of a ceasefire treaty in 1953, many villages or smaller towns would hastily be built close to US installations. They were to accommodate economically destitute Koreans who had come to these areas in search of opportunities in the shadow of the US bases that were destined to become permanent fixtures in the country even though the war had come to an end. Camptown areas— entertainment villages also known as *kijich'on* in Korean, or simply called "villes" among US soldiers—sprang up in many areas across the country; and it was Tongduch'ŏn, with particularly high numbers of US soldiers in the area, that would soon become the very synonym for these, at times, rather violent spaces of encounter between US military personnel and locals. The assemblage of land, property, and people under the auspices of the US Armed Forces, after all, meant a huge concentration of wealth in this predominantly rural area that had in the past only housed small-scale farmers and petty traders, and, in due time, it developed a strong gravitational force, drawing in locals in search of opportunities.

As a consequence, soon enough poverty-struck Koreans would come from different parts of the country—with a large portion of them being young women who followed the troops around to offer their sexual and emotional services. In this way, by the mid 1960s, Tongduch'ŏn had become home to about 7,000 prostitutes making a living from the soldiers (Moon 1997: 28)—a very high number, considering that the population of the town was 7,200 in the 1950s and reached 60,000 only toward the late 1960s (Kim B. 2007: 22). In addition to such rampant prostitution, the epidemic smuggling of PX[7] material into South Korea's black market, drug use, and other forms of illicit activities would flourish and soon turn this zone into a heavily stigmatized site where no ordinary Korean person would set foot.

Kijich'on areas nevertheless boomed during the 1960s and 1970s, when more than 20,000 Korean sex workers in total tended to the needs of about 60,000 troops in the country. To ease the extraction of US dollars that made their way into South Korea's economy, camptowns were now silently exempted from the Anti-Sexual Corruption Law of 1961 that prohibited sex for sale. Camptown areas were designated as "special districts" only a year later—districts in which "prostitution was not only allowed but was also closely monitored by the [Korean] Ministry of Internal Affairs, the Ministry of Health and Social Affairs, and the Ministry of Law" (Moon 2010c: 62). In order to regulate the women laboring near the bases, local and US military authorities would at times cooperate closely; for instance, mandatory STD (sexually transmitted disease) checks performed on the women were organized jointly (Moon 1997: 100; Lee N.Y. 2007), and direct nourishment of the sex industry seems to have been provided by Korean bureaucrats who delivered motivational speeches to sex workers in which they lauded them for their contribution to the local economy (Choe 2009; see also Lee J. 2010: 36; Moon 1997: 103).

But the 1980s already brought first signs of decline to the Tongduch'ŏn area: troop reductions (especially under President Nixon in the 1970s) and larger political, economic, and social changes in the country and the region deeply affected the camptowns near US bases (Moon 1997: 30–31). The dramatic economic ascent of South Korea that started to gain speed in the 1980s would further marginalize prostitution for US soldiers as a means of making a living, adding even more pressure to the women who already had to live with the mounting stigma attached to getting sexually involved with US soldiers. Consequently, many kijich'on sex workers now chose to move out of camptown areas to cater to a local clientele instead, which could be found in the now rapidly growing entertainment districts for domestic clients (Han 2001: 98–99).

Consequently, the Korean public interest in camptowns that was triggered by the murder of Yun Kŭm-i would, ironically, peak at a moment when those kijich'on spaces were already undergoing crucial transformations that would change these zones of contact for good. The news reporters, activists, researchers, and student protesters who sought temporarily to insert themselves into the world of kijich'on after the death of Yun Kŭm-i were chasing ghosts in more than just one sense: the booming of US military camptowns had peaked several decades before, when, under military dictator Park Chung-hee, any criticism of the security alliance with the United States, or the mere mention of rampant prostitution near US bases potentially came with a high price to be paid. In the early 1990s, however, the number of women employed in GI clubs had already drastically declined, and the many journalists who came to write on the plight of Korean camptown women often ended up pestering the same handful of women they still found hanging out nearby the bases (Han 2001; Moon 1997).

Figure 3.2   Quiet afternoon in a camptown north of Seoul

Camptown Fiction: Minjung Appropriations of US Entertainment Spaces

If camptown prostitution at the time of Yun's murder was no longer a social problem on the same scale as it had been in the 1970s, why exactly did violence against women perpetrated by US military personnel become such a pertinent public issue only after a 20-year delay, in the early 1990s? As I have laid out at length in the previous chapter, an explanatory model focusing on broader political factors would have to point toward the active dictatorial repression of any dissent against the US military presence from the 1950s until the 1980s on the one hand, and to the exponential growth of anti-Americanism among the dissident left of the country since the Kwangju uprising on the other.

When in May 1980, up to 2,000 protesters in the city of Kwangju were murdered by South Korean troops at the order of soon-to-be military dictator Chun Doo-hwan, the US Armed Forces stationed in the country were presumably letting the massacre unfold without interference—which came as a shock to many on the ground. "Given the privileged place of the United States on the cognitive map of South Koreans," argues historian Namhee Lee, "not only the US failure to intervene on behalf of the people but also its deep involvement in the suppression of the uprising was a rude awakening." She continues:

> The Gwangju Uprising proved decisively [to members of the minjung democratization movement] that the United States had not only been deeply involved in Korea but also had shared responsibility for the ugliness of Korean history, for its authoritarianism, military dictatorship, and political terror. (2007: 121)

With the rise of the minjung movement after Kwangju, therefore, gradually the realization started to sink in that the US was apparently not going to help facilitate the full package of democracy that it had promised South Koreans from the beginning. Opposition now quickly escalated to such a degree that it allowed what had previously been unthinkable: the denunciation of the older American brother who had seemingly turned out to be a false friend, and who, to make matters worse, could repeatedly be found going after the local man's women. Within such a politically heated atmosphere, it comes as no surprise that kijich'on—as symbolic and material terrains on the fringes of South Korea—would spark the imagination of quite a few writers and intellectuals from the minjung spectrum of society. Lurid accounts of camptowns and the life circumstances of the women employed in GI clubs, mostly in the form of fictive narratives, became a popular genre of writing among disenfranchised, predominantly male left-wing authors during the 1970s and 1980s. Much of this writing is filled to the brim with descriptions that deal with the issue of physical, cultural, and symbolic contamination through sex, in this way reminding us of Nira Yuval-Davis's insight that the "embodiment dimension of the racialized 'other' puts sexuality at the heart of the racialized imagery which projects dreams of forbidden pleasure and fears of impotency on the 'other'" (1997: 51).

Unequal power relations between men and women within a quasi-colonial context, Yuval-Davis further ascertains, are a potent breeding ground for imaginaries of sexual scenarios that mirror or invert the hegemonic order. A very real "absence of social responsibility toward the other [in colonial scenarios] often implies the freedom to violate and attack," she argues (1997: 52). In the Korean case, a state of near-immunity from local legal prosecution—provided for GIs by the Status of Forces Agreement between the two countries—for many decades entailed such an absence of responsibility. Bolstered by these legal provisions, and coupled with apparently widespread notions among the soldiers about the putative political, economic, and cultural inferiority of the locals, the young male strangers roaming the camptowns frequently engaged in petty crime and occasional graver offenses. These acts of violence, to connect Yuval-Davis's points back to David Graeber's *Dead Zones of the Imagination*, are then made sense of through "highly lopsided structures of the imagination" (2012: 119), with all the interpretative work falling onto the shoulders of the civilian population. The militarized male "others" who could be encountered by local Koreans in the entertainment areas—much empowered figures in the midst of at times highly disempowered locals— would now become imagined by actors of the Korean nationalist left *only* in the manifestation of potential rapist.

Outraged by stories of moral and sexual depravity that could be heard about kijich'on, the authors penning accounts of camptown proved to be the first disseminators of a kind of nationalist-driven violent imaginary with GIs at its center, an undertaking that initially came at a great personal cost for some of the authors involved: Nam Chŏng-hyŏn, for instance, who wrote his short story "Land of Excrement" in 1965, would be put on trial and go to prison for violating national security and anti-communism laws because of his fictional account of camptown (Hugh 2005). Literary scholar Lee Jin-kyung argues (2010: 134f):

> For its blunt opposition to US imperialism and militarism in South Korea, and its equally unambiguous use of a gendered and sexualized allegory, it would not be an exaggeration to say that "Land of Excrement" definitively established the genre of camp town literature at this particular historical moment.

"Land of Excrement" focuses on Hong Man-su, a black market dealer for PX goods, who avenges his mother's previous violation by raping the wife of Sergeant Speed. Speed, although not the GI responsible for the rape of Man-su's mother, has been sleeping with Man-su's sister, who is making a living as a prostitute in a GI club. Military prostitution, in this particular narrative, is only envisioned as "an institutionalization of the violence of imperial conquest as rape" (Lee J. 2010: 136), with the sexual encounter between the local woman and the foreign soldier being understood as always and necessarily steeped in violence.

After "Land of Excrement," several other authors affiliated with the minjung movement would make depictions of camptown a popular theme that allowed them to criticize the United States in the period of the 1960s to the late 1980s. In most of these writings known as camptown literature (*kijich'on munhak*), the explosive issue of US soldiers' sexual engagement with local women is seen as significantly connected to the perceived emasculation of male Koreans, with the biological, cultural, and symbolic reproduction of the nation seemingly gravely endangered by kijich'on practices of everyday miscegenation (Lee J. 2010: 125ff; Han 2001). The racial dimension of this putative humiliation of Korean men, while already playing a crucial role in the revenge-rape narrative of Nam Chŏng-hyŏn, becomes even more pronounced in "The Scream of a Yellow Dog" (1974), a short story by Ch'ŏn Sŭng-se. Here, another male protagonist makes his way into camptown to look for a young prostitute he knows, experiencing first-hand the depravity near US bases during his journey. At the end of the narrative, the protagonist and the woman watch a large white male dog rape a smaller yellow female dog, with the yellow dog finally letting out an eerie scream: "As Ŭn-ju and the protagonist watch the process, the man says to her tearfully, 'Ŭn-ju, yellow dogs must go with other yellow dogs, yellows with yellows'" (cited in Lee J. 2009: 140).

The themes put forward through camptown fiction did not stay contained within the written form; several movies have also been made about the putatively violent nature of the US–Korea encounter in kijich'on spaces. Ahn Junghyo's popular novel *Silver Stallion* (*Ŭnmanŭn oji annŭnda*), first published in 1986, was also turned into a successful movie in 1991, capturing new audiences with its story of how the rape of a young Korean widow by two soldiers triggers the moral destruction of a secluded mountain village that proves utterly unprepared for the arrival of thousands of GIs in its vicinity. And, in 2001, Korean director Kim Ki-dŏk would make the son of a black serviceman and his former prostitute mother, who keeps waiting for a letter from the US that never arrives, the main characters of his movie *Address Unknown* (*Such'wiin pulmyŏng*). An earlier movie on the subject is the 1988 film *Oh, Dream Country!* (*O, kkumŭi nara*), a film that also resulted in the filmmakers being taken to court and being fined after their movie's release (Gateward 2007: 206).

Oh, Dream Country! is particularly interesting in the way it shines a light on class issues in addition to the usual sexual and racial aspects of social relations in camptowns: the main character of the film is an unnamed college student who recently escaped from the bloody suppression in Kwangju, and is still haunted by guilt for having abandoned his working-class friends in that town to die on their own. Now, hiding out with a friend who happens to live in Tongduch'ŏn, the student becomes acquainted with several young Korean sex workers, all of whom seem deeply enchanted by the United States. A GI he befriends, at first a seemingly decent character, in the end commits an irreparable act of betrayal that ruins several lives. As America, the land of dreams and hopes, moves out of reach once again, the Korean characters are left behind in a world very reminiscent of Nam Chŏng-hyŏn's "Land of Excrement": while one sex worker who fell in love with the soldier kills herself, the protagonist's Korean friend murders the next American he randomly comes across. The protagonist himself, however, once again proves to be the most pathetic character when, startled by the events, he just goes on the run again, in this way evading his masculine (and national) duties once more.

Amplifying the Camptowns: Women's Bodies and National Boundaries

As we have seen in chapter 2, significant parts of the anti-US base movement that was to emerge with full force after the Yun murder came to be staffed by actors from the minjung faction—an assortment of left-wing forces that had traditionally been concerned with nationalist issues. Therefore, contentions over the woman's death were soon constructed by many voices of the movement as an issue pertaining to the fate of the nation, with alternative readings (such as feminist ones) quickly being relegated to the sidelines. In particular, it was the hegemonic understanding of the US military alliance as crucial for South Korea's security—supported by generations of conservative elites under the auspices of military rulers such as Park Chung-hee (1961–79), Chun Doo-hwan (1980–88), and now Roh Tae-woo (1988–92)—that was to be challenged by the promotion of a new focus on the violent nature of this partnership and the very real insecurity it brought to Korean women in their daily lives. Thus, within the explosive context of the presidency of Chun's hand-picked successor, former military man Roh Tae-woo, Yun's murder indeed proved to be the one event that would bind previously disparate political forces together into a new movement that could now formulate its grievances more sharply.

The process through which an apparently isolated event triggers large-scale structural—or even systemic—change has been named "structural amplification" by anthropologist Marshall Sahlins (2005). Seeking to describe "how small issues are turned into Big Events" (2005: 6), Sahlins explores pre-existing structural oppositions between different groups that become evident during conflicts, and how such differences are at times escalated into large-scale rows of a symbolic nature, especially if wider (imagined) communities such as classes, ethnicities, races, or nations become engaged: "upping the structural ante," he writes, "intensifies the battle, insofar as it now joins unconditional antipathies of morality and political ideology, not to say cosmology, to petty disputes that otherwise would be negotiable" (2005: 25). Utilizing this notion for the events of 1992, the Yun Kŭm-i murder can thus be read as an example of how the conversion of a micro-history (a fight between a sex worker and her client that ended in a gruesome act of violence) into a macro-narrative that pertains to the fate of greater collectives (the potential death of the Korean nation at the hands of America) may rapidly unfold in a volatile political climate of change. "Collective subjects such as nations," writes Sahlins, "'imagined' as they may be, take on the flesh-and-blood qualities of real-life subjects [...] and are accordingly acted out in interpersonal dramas, with all their attendant feelings and emotions" (2005: 6).

A double amplification in fact took place during the protests following Yun's death: specific places (that is, the kijich'on areas adjacent to US military bases) were now turned into imagined spaces that were understood by a great many people not directly acquainted with them as both national spaces of shame and transnational spaces of empire, where US domination over South Korea materializes in its most violent manifestation. In contrast to David Graeber's bureaucratic spaces, which he describes as dead zones, these camptowns proved rather to be vital zones of the imagination for those concerned about their existence. Partly as a result of depictions of death and violence in these US military terrains, camptowns have paradoxically become dynamic zones that played a generative role for those trying to contest the US–South Korea military alliance. The appalling killing of Yun, one single "critical event" (Das 1997), in this way came to highlight pre-existing structural conditions of inequality that proved to be particularly well-suited for a process of amplification. Consequently, camptowns were turned into a symbol of Korea's suffering as a nation, and the women living and laboring in them were for a short while typecast as the long-lost daughters who had been sacrificed by the South Korean conservative elite in the name of the country's security.

Building on widespread colloquial notions that these areas were de facto non-Korean territory, where no "decent" Korean woman would want to go, the heavy weight of the stigma of disreputability that the sex workers laboring in these areas were suffering from was all the while *not* being neutralized by such attempts to incorporate "the" camptown prostitute into a nationalist framework. As writers such as George Mosse (1988) and Sherry Ortner (1978) have shown, women tend to figure in the symbolic make-up of larger structures such as the nation-state predominantly as subjects whose sexual purity is in need of policing, with virtues such as female respectability, a "good" moral standing, and sexual propriety adamantly promoted by those elites seeking to further their nationalist projects. While at first glance it might seem subversive that a sex worker should become an icon and stand-in for the putatively violent relationship between the United States and Korea, upon closer examination we see that traditional notions of female respectability were not in the least bit undermined by such an appropriation. With "the" prostitute consistently envisioned as a social type who exists outside of the realm of good female behavior, she is also understood as positioned on the very margin of the national community, at times being considered a threat to its integrity and moral superiority. If the camptown woman, through the posthumous appraisal of Ms. Yun, was now to be included in the realm of the nation, this could only occur by firmly casting her in the role of hapless victim, and victim only. In this way, the old binary opposition of the prostitute as either victim or villain was finding another kind of re-appraisal in the aftermath of Yun's murder that was also bolstered by the involved feminist activists' anti-prostitution agendas.[8]

In this way, stories of sexual exploitation and violation of Korean sex workers in the entertainment areas near US bases were at times used as all-too-neat allegories for the suffering of the Korean nation as a whole. Thus the plight of Yun Kŭm-i came to support the discursive construction of the nation in the early 1990s, a nation that was understood to be under permanent duress from attacks of malevolent outside forces (first by Japan, now by the US). And while the imagined community of Koreans under military dictators Park Chung-hee (1961–79) and Chun Doo-hwan (1980–88) was mostly promoted as a nation of soldiers, as Sheila Jager has explored (2003), now a new imagination was given more space among left-leaning nationalists: that of the nation being likened to a woman in duress.[9] However, ultimately such an understanding of the nation may have been rooted as much in a firmly patriarchal worldview as that of the soldier-nation, with the women unwittingly becoming:

> both the object of concern over the state of Korea's racial (inner) "purity" and the subject of active resistance to (outside) foreign "contamination". [...] The threat to feminine chastity (and by extension, to marriage) was perceived as a threat to the integrity and "inner" (racial) continuity of the nation itself. (Jager 2003: 73)

Furthermore, the appropriation of Yun Kŭm-i by the nationalist-leftist movement ultimately allowed very little space for the particular personal details, histories, and motifs of the women whose narratives were being used. With Yun Kŭm-i no longer being able to speak for herself, other contradictory voices had to be kept out as well, such as that of Kim Yŏn-ja, the first former kijich'on sex worker to openly talk about her experiences through the publication of her autobiography *America Town*. She claims that:

> there were dozens of girls who died before Yoon Geum-yi [i.e. Yun Kŭm-i] died. But no one ever tried to help us when we called for help. [...] I felt that Yoon Geum-yi was just used as a tool for anti-American protests. (Lee M. 2005)

In a similar vein, Han Jung-hwa, a German-Korean scholar-activist who studied the camptown areas in the early 1990s, claims that:

> if one were to take the personal decisions made by the women seriously, one would no longer be able to co-opt them as national victims. The women are in fact more victims of [Korean] society's moral double standard and of nationalism. (2001: 99, my translation)

Derogatorily labeled by others as "Yanggongju" (Western princesses) or "Yangkalbo" (Western whore), camptown women have in the decades since sought to counter their narrow role as silent victims by asserting their key role in keeping the boundary between Korean nationals and GIs intact. Tapping into the nationalist concepts they have repeatedly been subjected to by both the South Korean military regime and its civilian opponents, they assert the importance of their role for the country's economy and security. For instance, one camptown woman, in an interview with a reporter from *Mal* (a monthly left-leaning magazine), argued that:

> They [i.e. the GIs] rape even when there is prostitution. What would it be like if there were no prostitution? Wouldn't the rape of our country's women in the surroundings of military bases increase? We are the breakwater (*bangpajae*) that stops these things from happening. We should not be despised for what we are doing. (quoted in Cheng 2010: 68)

Such individual attempts at asserting their own value are on occasion augmented by more collective endeavors to restore the women's standing in society, which recently culminated in a lawsuit, filed against the Korean government on July 25, 2014, which is still ongoing. The 122 female claimants, supported by a network of NGOs, have argued that the government should compensate them for the past injustices they encountered because of the work they engaged in on behalf of the Korean state: "They say we were patriots at the time, but now they couldn't care less," former prostitute Kim Sook-ja summed up her viewpoint on the matter in a Reuters news article: "We didn't fight with guns or bayonets but we worked for the country and earned dollars" (Park J. 2014).

The Stigma of Miscegenation

Stigma, Erving Goffman has famously argued, is "an attribute that is deeply discrediting" (1990 [1963]: 13) and that reduces the bearer "from a whole and usual person to a tainted, discounted one" (1990 [1963]: 12; see also Link and Phelan 2001). Despite their attempts to claim a higher public standing (as discussed above), the stigma from which Korean camptown women suffer still weighs heavily indeed. In large part, this widespread disregard stems from their sexual involvement with US troops, which has diminished their worth in the eyes of ordinary Koreans. Their sexual activities with foreigners, so goes the all-too-familiar narrative, have quite literally exposed the Korean nation to the greatest threat: that of miscegenation. The sex workers are often understood to be the very embodiment of that peril of treacherous mixing, and this is particularly true for those women who have given birth to children fathered by US soldiers. With these women's life-narrations exemplifying the dangerous hybridity that may emerge between Korea and the US if local women go with US soldiers, their "mixed-blood" children are often seen as living proof of their "tainting" encounters with the soldiers.

Up to 100,000 women have migrated to the US as so-called "military brides" since the Korean War.[10] Anecdotal evidence seems to indicate that broken marriages among the "military brides" are very widespread—many stories of desertion, mental breakdown, suicide, and homelessness among these women haunt the Korean-American diaspora, as Grace M. Cho, in particular, has explored (Cho 2008; see also Chung n.d.; Yuh 2002).[11] Some of these women, after falling on hard times, have returned to South Korea and have ventured back to the camptowns, that is, to the only spaces that are truly available to them after their life experiences have placed them so far outside of the collective imaginary of what it means to be a "decent" Korean woman. Here, they join other aging Korean females who have never managed to leave for the States in the first place, and who often reside in camptowns because these are literally the only areas left for them to go to, with their friends and families often having cut ties with them years or even decades ago (Moon 1997).

Their children, if abandoned by their American fathers, frequently experience discrimination on a daily basis in Korean schools, in the neighborhoods, or at the workplaces they labor in (Lee M. 2008). Many of the women sought to prevent such a fate of exclusion from Korean society by sending their children away for overseas adoption. Accurate numbers of how many children of camptown sex workers were adopted are not readily available. But between 1955 and 2000, approximately 140,000 children, most of them born out of wedlock, were sent abroad to be adopted by new families. Initially, Eurasian children that were the outcome of liaisons between GIs and Korean women formed the majority of these children, with their share only decreasing in these transnational adoption flows once the US troop numbers in South Korea were significantly reduced in the 1970s (Lankov 2007: 252f).

Moreover, even women whose children were fathered by Korean men may at times find their families ostracized from Korean society if they stick around in the camptown areas for too long, as I was to learn one afternoon during a visit to a camptown near Ŭijŏngbu. Mrs. Kim is a woman in her 40s who works as a waitress at a GI club in the area. I met her and her 17-year-old daughter at a little burger joint close to Camp Stanley's side gate.

Mrs. Kim came to this town ten years ago, she tells us; she ran away from her husband, and took her two children along. For five years now, she has been working at a GI club. "It's ok work," she says, and adds that she would like to stay in this town. "I have nowhere else to go, even if the base were to close.... It's just that my children hate it here so much ..." She is very worried about their future. "I keep telling them, after they turn 20, there's nothing more I can do for them. I can't pay for their college, I can't help them out at all, and I want to live a bit myself still, too, you know." Her son is 16 now, and both he and his sister hate even to leave the house in the evenings. "Are they afraid of the GIs?" I ask. "Not afraid, no, they just hate it, they hate everyone, they think we are all gangsters." She further explains:

> It's my in-laws, they keep telling the boy things like that. In the movies, they always see that the gangsters smoke and spit and sit around at the clubs all day, and then my son looks at us, and he says, "Mama, you're a gangster, too, no?"

The owner of the burger place we were in, a Korean woman who looks like she is in her 60s, told us that she used to be a real beauty back in the day, and an actress on top of it, "but look at me now," she said and laughed, returning behind the counter to make more burgers for us. Her eatery is small and dilapidated; the dark furniture takes the visitor straight back to the 1980s, the chairs and tables look as if they might not last the day, with one lonely fan swirling up the dust rather than cooling the place down. The only decorations I could make out on the wall were two framed black-and-white photographs of a curvy American actress, in a racy pose. Next to these, I can see a series of snapshots of an African-American GI: "Funny guy. Came here all the time, but he's in Iraq now," the owner shouts above the noise of the ventilator. At that moment, another soldier, perhaps 20 years of age, steps in to order some food, which he wants the mama to deliver to a nearby club. "Ok baby," the owner says, and when the guy leaves the restaurant, we clearly hear him yell down the street at a Filipina woman passing by: "Hey, bitch! Hey! You fucking bitch!" Mrs. Kim's teenage daughter asks her mother what the serviceman had just said. Her mother just smiles at me and then says in Korean, "No idea what he said, no idea." "Whenever they swear," she adds in English so her daughter does not hear, "I just don't understand ..."

Remembering Yun Kŭm-i, Forgetting Her Sisters?

"Forgetfulness," French historian Ernest Renan once wrote, "form[s] an essential factor in the creation of a nation" (2001: 166). But rather than willful forgetting, as this particular moment in Korea's history shows us, the more active moment of *co-optation* of marginalized voices has proven to be even more crucial to this nationalist project, seeking to make its vision of the imagined community the dominant narrative. Furthermore, the women whose bodies, sexual labor, and lived experiences were being dissected and re-assembled in such a way, would often find only increased stigmatization and marginalization down the road, with the very spaces they had come to inhabit now becoming endangered on top of already being endangering.

In the meantime, new levels of prosperity have vastly changed the reclusive camptown spaces: instead of local women, nowadays it is foreign sex workers who labor in the GI clubs—a phenomenon that is symptomatic of the dramatic shift in the role South Korea plays in the world economy today, with the country increasingly opening up its own frontiers of capitalist exploitation for itself within the wider region, and extracting surplus labor from migrants for whom South Korea has become an attractive destination (see, for instance, Kim A.E. 2008; Park W. 2002). Consequently, hyper-marginalized female migrant workers from the Philippines and other countries in the wider region nowadays mingle in camptowns with American soldiers of various ethnic and social backgrounds. Due to the ever decreasing spending power of these young men as compared to the host population and their symbolic devaluation since the democratization process of the 1990s, the soldiers often understand themselves to be just as ostracized in modern Korea as their "third-country national" entertainer counterparts—a dynamic that will be further explored in the next chapter.

To be sure, in comparison to the heyday of kijich'on in the 1960s and 1970s, the number of sex workers employed in camptowns nowadays has drastically decreased: in 2009, for instance, around 2,300 women came into the country with the help of entertainment (E6) visas (cf. Rabiroff 2009), the majority of whom would have been destined to head for camptown clubs and bars. The voices of these new women in camptowns, perhaps not surprisingly, are rarely ever heard in Korean public discourses and kiji'chon spaces are nowadays increasingly forgotten again, with even most of the nationalist groups that used to be outraged by the social conditions in these areas during Yun Kŭm-i's time no longer paying them any attention. In contrast to the older Korean "yanggongju," after all, these female strangers stand unambiguously outside of the imagined moral circle, the boundaries of which are defined by membership in the Korean nation. The fact that it is primarily Korean club owners and local madams who make use of these women's sexual labor to siphon off much of their hard-earned money into their own pockets is seemingly not an issue that causes much outrage.

When camptown spaces first came into view in Korean public debates as violent spaces, from 1992 onwards, this kind of imagination gave local nationalist actors an excellent tool that allowed for a drastic repositioning of South Korea in relation to its big ally. With the symbolic struggle that broke out over a space that tends to be erased from visibility in Korea over and over again, the neighborhoods at the center of attention were—ironically—driven even further into marginalization. The appropriation of the Yun murder, we have seen, proved significant in popularizing the image of US soldiers as violent brutes and possible sex offenders on the loose in the remote adult entertainment spaces near their military bases. The aftermath of the event, as I have shown, also needs to be understood within a longer-term project of actors from the leftist-nationalist end of the political spectrum who attempted to symbolically recalibrate the hierarchical relation between the two countries: an image of brotherly affection that was a common trope to depict the relationship between the two nations was now partly eroded by the painful realization that the powerful American friend could regularly be found going after local men's women. Related to this, the possibility has to be considered that the figure of the camptown woman, that is, the Korean sex worker employed near the bases, was utilized as a symbol of the nation under duress by nationalist actors of the left in the country precisely because it conjures up widespread familial and sexual anxieties among (male) citizens of the nation. The emotive strength of such a bodily image of the nation as a ravaged woman lies precisely in the fact that it has proven repeatedly to be a trope that is very easy to think and act with during times of political upheaval.

4

Vil(l)e Encounters
Transnational Militarized Entertainment Areas on the Fringes of Korea

In the Shadow of the Base

Rose is sitting in front of her little diner together with a few of her friends, and she invites me to have a coffee with them. Rose is a Korean woman in her mid 30s, but looks at least 10 years older than her actual age—a result of all the heavy drinking that her former work at the GI clubs nearby entailed. Nowadays she runs this small eatery in the ville, but things are not going too well for this new business of hers, either. Many of the soldiers in the area have recently been relocated to P'yŏngt'aek, which has hit the remaining residents in this kijich'on North of Seoul rather hard. This is why she had to close down the little club she was running up until three years ago, she tells us. Even though the diner that she owns now is not doing any better either, Rose says she still prefers GI customers over Koreans: "The Koreans, they think they are the king, and they treat you like a slave, while the GIs, they just come, they want one big meal, no fuss, no extra wishes, no bullshit."

If the US base Rose lives alongside of were to close for good, maybe she should seize the opportunity, she says, and "get the hell out of here, go to the States myself." She has been contemplating such a move also for the sake of her son—he is 16 now and, with her failing business, she is very concerned about his future prospects. Being the offspring of a soldier, he has been attending US military schools for most of his life. And although he can read, write and speak in Korean, he is at a great disadvantage compared to his age cohort when it comes to passing the necessary exams that would allow him to get into a Korean university. Rose's son causes her a lot of trouble these days "with all his running around," she says, but the fact that she can occasionally speak with the boy's grandfather in the States on the phone gives her some consolation. Grandfather keeps telling

Figure 4.1   Fenced-off US military installation

her that "Kids will be kids, give the boy some time, things will change." With the actual father of her son, however, she barely has any contact at all—her ex-partner does not send him any gifts, not even a card on birthdays or Christmas. "Asshole," she concludes the talk about him, "he has so much more money than I do, but still, he doesn't do anything for his son … You know what my kid said to me the other day? 'Basically I have no father, right?'"

Miss Yu, another woman in her early 30s, jumps in now, telling us of her work at a club nearby where she is employed as an entertainer. She finds the younger Filipina women that she has to work with quite irritating. They rarely talk to her, and even when they greet her, she says, they do it in such a mocking fashion that she feels rather insulted by them. She has this distinct feeling that these Filipinas are gossiping about her behind her back, constantly switching to Tagalog when they are in the midst of a conversation, and giggling obnoxiously at the same time. At least there is Natalia, a Russian woman who is 22 years of age. But then Natalia has a problem with alcohol and, whenever she drinks too much, she gets very aggressive, swearing at everyone in Russian and being abusive toward the girls she does not like. On one occasion, she got herself into a fight with a Mongolian entertainer, tearing out a lot of the woman's hair. "So whenever Natalia gets too drunk," Miss Yu says, "I just make sure that the girl gets a bit of sleep until she is fit enough to work again." A reason for all the tensions amongst the women working at the club, Miss Yu believes, is that everyone has been so stressed out about money lately. Barely a customer has made his way to their club over the last few weeks, and the financial situation of the business seems to be getting worse by the day.

In the last chapter, we saw how various symbolic struggles in South Korea that became more predominant in the 1990s put camptown areas into the national spotlight, where they served as vital zones of the imagination during a tumultuous political era. Gruesome depictions of camptowns have accumulated through a social practice that I call violent imaginaries, which, as I have explicated in the introduction, involves the rescaling of individual acts of violence into a matter that pertains to the Korean nation. But these images, accrued via a political project that is aimed at shifting larger US–Korea relations, do not entirely capture the reality of contemporary camptowns as I experienced them in the late 2000s. Rather than finding contested zones where Korean women are routinely mistreated by US soldiers—an image of camptowns that spoke well to the national imagination of the 1980s and 1990s—I discovered a number of marginalized transnational spaces where a diverse set of actors encounter each other in a very particular landscape.

These meetings, while potentially open-ended, are in practice often scripted by the fact that this assortment of soldiers, foreign entertainers and local women confront each other within an economic micro-system that largely encourages sex for sale, with the Korean club owners, in particular, seeking to sideline all other, non-monetary relations between women and soldiers that they are unable to make a profit on. This order regulating the entertainment areas, which places sexualized entertainment squarely at the center of all economic activities, in practice often sidelines the Korean women still working in camptowns, who find themselves unable to compete against the young female foreigners in their midst. With a woman's worth largely defined by her ability to attract the attention of the American customers who come to the club she works at, young and physically attractive entertainers are typically in a much better position to make a living than are women who are older. This distinction between younger and older women, between potentially high-performing entertainers and those who need to content themselves with meager wages, is usually drawn along ethnic lines in the camptown areas I visited in 2009:[1] most of the young entertainers are from the Philippines or Central Asia nowadays, while the Korean women—typically in their

30s, 40s, or older—have often quit club work in order to take up other positions in the area, such as managing the clubs' daily affairs, waiting on tables, cleaning, bar-tending, etc. If they do become too old to perform even these low-paid tasks, they may come to depend on the (very minimal) welfare support provided by the Korean state, or on donations made by Christian groups or non-governmental organizations (NGOs).

As for the foreign entertainers themselves, the business with sex is not necessarily a lucrative one for them, either. The women usually receive a base salary of up to $400 a month; the rest of their money they have to earn through commissions they receive on the overpriced drinks they sell to the soldiers[2] in exchange for a little bit of their time at the club. There is also the extra income that can be made by going on "bar fine" with a client. The bar fine is a fixed amount of money (usually at least $100) that needs to be paid to the club owners in order to be allowed to take an entertainer outside of the bar for the remainder of the night. The entertainer herself typically only receives a fraction of that sum, which is why many of the women seek to arrange their own meetings with clients without the knowledge of their club owners. De facto, the bar fine is often the equivalent of a sex fee; but the term itself and the conditions of this arrangement are left vague enough to absolve the club owners from accusations of facilitating prostitution, while most of the weight of any illicit transactions is placed on the shoulders of the entertainer, who putatively has to decide for herself how far she wants to go with her customers. In reality, however, soft and sometimes outright pressure to perform sexual services on a night out is often exerted by the club managers or the soldier who has paid a lot of money to be in the company of the woman.

While all these monetary and sexual activities are going on, at the same time many of the daily goings-on in camptown revolve around the women's efforts to re-embed their brief encounters with the soldiers into the non-monetary realm of social obligation, longer-term commitment, and possibly even love. The women engage in myriad strategies that are aimed at disrupting the strictly transactional nature of their relationships with the men who come to their clubs. In doing so, they often resemble their Korean predecessors at the clubs, who frequently worked at these establishments with the stated goal of finding themselves an American husband. And some of the tensions between the older Korean women in camptown and the younger foreign women at the clubs may well be grounded in the fact that the Koreans typically look back on a long line of failed relationships with soldiers, with their chances of finding "a good guy" rapidly fading over the years, while the younger foreigners usually still have the highest hopes for themselves. Suzie, a Korean woman in her mid 30s, whom I often found in a somewhat intoxicated state, sitting at this bar or that while loudly complaining about the Filipina women in town, summed up this key dilemma of her life in the following way: "Yeah, sure, I used to go with all the guys. And boy, did they treat me badly. I am done with all that now. Just want a husband now, and want to be loved."

Camptown Residents: Hopeful Actors or Preoccupied Persons?

In an attempt to navigate the social stigma attached to their line of work, the Filipina women who perform the actual sexual(ized) labour often place their bets entirely on their soldier-clients when it comes to improving their chances in life. Sallie Yea, who conducted a series of interviews with Filipina entertainers in South Korea in 2002/3, has called the romantic underpinning through which the women make sense of their work a kind of "labour of love." Such a romantically charged framing of their lives, Yea argues, allows the migrant women to assert a degree of agency while working in this highly unfavorable environment, and to temporarily overcome the stigma attached to prostitution (2005: 457). Sealing Cheng, in her ethnography *On the Move for Love* (2010), advances a similar argument when she notes that the "vitality of hope [...] propels these women's transnational movements" (2010: 222). In the foreign women's attempts to make their current lives in camptown meaningful by imagining a brighter future ahead, GI boyfriends play a vital role. Soldiers, Cheng notes, emerge as the women's "prime concern, and preferable source of support" (2010: 6):

> This strange intimacy between the Filipinas and the GIs in the context of *gijichon* is a main theme [...]—"strange" because of the dominant/ subordinate political, economic, and historical relationship between them, a structural relationship that has predominated in discourses about the US military and women in rest and recreation (R & R) facilities. (2010: 6)

Such a focus on hope also reflects a wider trend in social anthropology, where the term has recently gained much traction (for a review, see Narotzky and Besnier 2014)—Hiro Miyazaki (2004, 2006), for instance, has argued for a deeper exploration into the ways that hope figures in various forms of knowledge production. Vincent Crapanzano (2004), on the other hand, speaks of our informants' "imaginative horizons" as a potentially fruitful terrain for ethnographic research, while Arjun Appadurai has advocated for a "politics of hope," which people engage in in order to transform uncertainty into a manageable risk (2013: 115ff). And Frances Pine, who explicitly links migratory trajectories to hope, explains that "hope is a complex, many-layered notion resting on the capacity for imagination, on a sense of time and of temporal progress, on a desire to believe in a better future or in the possibility that something can change" (2014: 596).

Analytical approaches that seek to locate these migrant women's agency in their ability to manage their daily lives in kijich'on through their hopes of a better future are unquestionably important contributions to the larger victimhood vs. agency debate that has been dominating research on prostitution / sex work for decades now (see, for instance, Agustin 2007; Berman 2003; Day 2007; Day and Ward 2004; Doezma 1998; Kempadoo 2005; Kempadoo and Doezma 1998; Weitzer 2000, 2005).[3] The term *prostitution* notoriously entails an impossibly diverse range of activities and systems across the globe (Donovan and Harcourt 2005; see also Majic 2014), with the case of Korean camptowns certainly having much to contribute to these larger discussions. However, my goal here is not primarily to contribute to this body of literature, but to discern what is arguably the unique feature of the kijich'on system—that is, the fact that entertainers and soldier-clients encounter each other within a highly militarized environment, and find themselves drawn into a system that appropriates their labor for the benefit of a security system that spans half the earth. While I fully support a project that accentuates the agency of sex workers, paying close attention to the larger structural forces that these actors need to come to terms with is also of crucial importance.

By keeping the sweeping historical forces in mind that have given rise to these red-light districts adjacent to US bases in South Korea, when thinking of the women and men who encounter each other in camptowns, another question may arise: Can we actively bring militarism into the equation again, while at the same time avoiding the pitfall of portraying the women involved in camptowns as mere victims? And if outright violence does indeed play less of a role in the maintenance of the camptown system than nationalist actors in Korea would have it, what other, softer forces keep these kijich'on actors in their place? While I believe that the rather generic term "hope" may not provide a satisfactory answer to this type of concern, it does lead us in a very interesting direction—that is, to the terrain where personal aspirations, collective imaginaries, and various temporal orientations come up against a local architecture (i.e. the world of kijich'on) that has attached itself to the globe-spanning infrastructures of the US Armed Forces.

As discussed in chapter 1, militarism is perhaps best understood not only as an ideological matter that pertains to how military practices are made sense of but it is also a phenomenon that has significant, yet often unacknowledged effects on actual social relations. Militarism, whether explicitly verbalized or not, often permeates the everyday lives of civilians in numerous ways. In South Korea's US camptown areas, then, we may find ourselves on a rich terrain to explore the impact that one particular military has had on sexual and romantic encounters in its sphere of influence. Sandya Hewamanne, in her work on Sri Lankan soldiers and their girlfriends, has proposed the term "preoccupation" to capture a sense of the extensive emotional work that goes into maintaining sexual and romantic relations in highly militarized environments. Militarism, she has shown, indeed did not remain at the discursive level of nationalist ideologies at her field site, but tended to seep "into the intimate, everyday spaces of supposedly peaceful areas" (Hewamanne 2013: 61) among working-class women in southern Sri Lanka, who typically forged sexual liaisons with lower-ranking soldiers. Describing these female factory workers she encountered as "friendly fire casualties of war" (2013 62ff; see also Lutz 2002b: 88), Hewamanne looks into how "*occupation* can take many affective forms, sometimes entailing pleasure, pride, hope, and opportunity along with pain, fear, and violence" (2013: 66). The women she lived with did indeed invest much emotional labour in their soldier boyfriends, albeit often with little actual success: "Occasionally, one heard of soldiers marrying their [...] worker girlfriends, but mostly they were said to cheat, abuse, and abandon them" (2013: 67).

"Preoccupation," laid out in such a way, seems to be an excellent phrase to also capture the intense emotional and sexual involvement that women working in camptowns attempt to establish and maintain with the US soldiers they meet in the ville. While "preoccupation" today is most commonly used to refer to a mental state of near-total absorption, the term has military connotations as well: it is derived from the Latin word *praeoccupare*, that is, "being seized beforehand." Additionally, it points to people's active engagements with their futures: during the 16th century, for instance, the word was still used in the sense of "meeting objections beforehand."[4] By keeping these two older meanings of "preoccupation" in mind, we may be able to make this term speak to a wide emotional spectrum that goes into the sexualized labour that the women are engaged in. While "hope," in a sense, primarily points to those sentiments and feelings that are more firmly anchored in individual choices and aspirations, "preoccupation," in my understanding, is much more tied up with the notion of affect—that is, it highlights how people's emotional agitations are both a deeply embodied experience *and* a phenomenon that significantly emerges in the collective, in the encounter with those who are (not) like us (Mazzarella 2009, 2015).

And, indeed, the sensual and emotional underpinning of camptown life, the sum of these countless interactions between soldiers and women that play themselves out every night, may in fact be the "soft" stuff that ultimately keeps the uneven system of the camptown in place. If the ville is a pressure valve for the soldiers who visit it in their free time, a place where they can "let off some steam"[5] after the day's work is done, it is simultaneously also a space of desire, uncertainty, anguish, and hope for the women who labor there. Affect, William Mazzarella has claimed, is a crucial component of structures of power in the sense that "any social project that is not imposed through force alone must be affective in order to be effective" (2009: 299).[6] If we take this explosive claim seriously, then camptowns could simultaneously be understood as a historical project arising out of Cold War insecurities, an economic and political solution to the perceived danger posed by too many young foreign men in one place, and a social project in Mazzarella's sense that is affective in its effectiveness as a reality to be lived in by those who temporarily find themselves in it.

What about the soldiers then? Are they equally "preoccupied" with the entertainers they meet in the camptowns close to their military installations? While the young women's lives were typically dominated by their involvement with a number of soldiers—which, in addition to the sexual(ized) labor of the night, also involved grooming and beautifying oneself during daytime, as well as engaging in the almost endless work of texting (potential) clients during the day—the soldiers I spoke with were often also quite invested in the emotional work of building relations in camptown. With kijich'on areas functioning as temporary escape zones that allow the soldiers to relieve the stress of the demanding work they engage in, camptowns certainly become emotionally charged spaces for many of the servicemen, too. Their "preoccupation" with the relations forged in these spaces, however, is often of a qualitatively different kind than that of the women, given that their monetary interests are essentially opposed to those of the entertainers.

The fact that money plays a crucial role in the way soldiers and women first encounter each other certainly makes the emergence of romantic relationships difficult, to say the least, with the "whore" stigma that the young entertainers seek to manage also directly affecting their interactions with the men they meet in the clubs. The soldiers, some of the women complained to me, typically seemed to assume from the start that the entertainers they encountered in the clubs were only after their hard-earned money, and thus frequently treated them with little respect. All the while, however, quite a few soldiers were receptive enough to the presence of these women, and forged (temporary) relationships with these other working-class strangers living and laboring on the outskirts of Seoul. Within this context, some soldiers opted for a very different evaluation of these entertainers; they, too, chose to portray the women as victims who are involuntarily trapped in a vicious prostitution system. Tony, a 26-year-old soldier who had been stationed in South Korea for four years, explained to me one evening that much of the bad reputation of the US military in Korea stems from the women present in the villes who are "brought in for specific purposes," as he put it. In his hometown in New Mexico, there were also many female Philippine residents to be found, he

Figure 4.2   A group of servicemen in a kijich'on

said. "They are beautiful, beautiful women. I just love them." Recently, he re-visited Tongduch'ŏn's ville for the first time in a while. "I met a very nice Filipina lady there in one of the clubs," he explained. Instead of paying the bar fine for her to purchase her company for the night, he asked her out on a date, but she told him that the club owner would probably not allow her to do that. "You know what that means, right? She's basically being held there like a slave. Many are. It's sickening."

Transnational Migration Circuits Into the Entertainment Industry and Debates on Sex Trafficking

> Lana came to South Korea for the money. Back home in the Kyrgyz Republic, where she toiled in a shoe factory for $20 a month, she longed to buy an apartment, but the $5,000 price tag seemed impossibly high.
>
> Then she saw a newspaper ad seeking women to dance and talk with U.S. servicemen in nightclubs in South Korea. The ad promised what for her was an astounding wage—$2,000 in the first six months. Lana, a bright, attractive blond, took the job.
>
> Now, she wishes she hadn't. (McMichael 2002)

In the early 2000s, a number of articles were published in American media outlets concerning the issue of sex trafficking of women from post-Soviet areas and the Philippines to kijich'on terrains in South Korea.[7] Tom Merriman, a US reporter working for the Fox television affiliate WJW, got the ball rolling on this story after he had encountered a number of Korean sex workers in Cleveland's massage parlors. These women, he learned, had ended up in the area because they had typically followed US servicemen to the United States, where they then resorted to prostituting themselves after their relationships with these men had ended. "I always had an interest in how these women got here," Merriman told another journalist (Jacoby 2002), and he would consequently travel to South Korea's camptowns in a quest to learn more about the matter. To his surprise, though, in the meantime the Korean women he expected to find nearby US bases had largely been replaced by Russians and Filipina entertainers. The reporting that followed after Merriman's television station aired secretly filmed footage of US soldiers pursuing foreign entertainers generally focused on how these women had allegedly been lured to Korea under false pretenses, where they were then pressured or directly coerced into prostitution. An article entitled "Sex Slaves" (McMichael 2002; see also excerpt above) gives a good indication of the kind of reporting that was done on this topic of female migrant workers who came to South Korea on "entertainer" visas to work in the GI clubs nearby US bases.

While recruiters—who often use illicit practices to get women to sign up for this work—certainly played a significant role in bringing Russian entertainers to South Korea, some of the migratory paths of these young women were most likely also facilitated through informal networks among the sizable migrant community of Russians living in South Korea. Two Russian women, to whom I spoke during field research in It'aewŏn, essentially told me that going to South Korea to work at a club was thought of as a good option to make some quick money among their circle of female friends. Unemployment rates are incredibly high in her native town of Vladivostok, Alexandra, a 24-year-old woman said. Many of the young people dream of finding jobs in Japan or South Korea, places which to them represent much more attractive destinations than the far-away urban centers of their own country. Yuliana, a 26-year-old resident of It'aewŏn who also hails from Vladivostok, says she came to Korea for the first time at the age of 21. She recently got married to a South Korean man, but continues to work part-time at a Russian club in It'aewŏn. Her dream is to go into retail and eventually to establish a little trading company for herself. This would allow her to ship goods to her hometown of Vladivostok. Since the late 1980s, Russians have done exactly that: attracted by Seoul's Tongdaemun area, the old heart of South Korea's garment industry that is still a regional center for the trading of cheap clothes, many have come to South Korea to try their luck in small-scale retailing. An estimated 70,000 Russians, quite a few of them undocumented, were residing in the urban centers of South Korea until a major crackdown against illegal immigration took place in late 2003/ early 2004 (Chun 2004). As of 2013, 12,800 Russians still live in South Korea legally (5,000 of whom are made up of ethnic Koreans with Russian citizenship), with the number of undocumented Russians being estimated to range in the thousands as well.[8]

At approximately the same time that Russians in South Korea experienced this crackdown, South Korean NGOs and women's right groups were increasing the pressure on their government to take the issue of sex trafficking more seriously (Moon 2010a: 347). Curiously, they found their anti-prostitution stance supported by an unexpected ally: the same United States government whose soldiers happened to be the main clients for the Filipina and post-Soviet women working in clubs close to US military bases. Already in the year 2000, the US government had introduced its

Trafficking Victims Protection Act, and would consequently attempt to globally combat prostitution (which was identified as a prevalent form of trafficking) through various ways and means (Cheng 2015). For instance, it began to utilize the annually published *Trafficking in Persons Report*. By assessing and grouping various countries' anti-trafficking measures into three tiers, this report proved to be a vital tool in shaming states into cooperation with the US anti-prostitution agenda. In 2001, South Korea was listed as "tier 3" in one such report. That is, it was named to be among those countries whose governments do not fully comply with the minimum standards and are not making significant efforts to do so. While in that same report, South Korea had only been listed as a source and transit country for human trafficking, by 2002, it was also named as a destination country, with "persons from the Philippines, China, Southeast Asian countries, Russia and other countries of the former Soviet Union" named as those who are primarily trafficked to Korea.[9] This kind of exposure seemed to have played quite a significant role in how swiftly South Korea came to replace the old Prevention of Prostitution Act with the Act on the Punishment of Procuring Prostitution and Associated Acts and the Act on the Prevention of Prostitution and Protection of Victims Thereof in 2004 (Cheng 2015).

Another response to this rather unwelcome international attention had already come a year earlier, in 2003, when the Korean government stopped issuing E-6 visas to women from Russia and other post-Soviet spaces (Moon 2010a: 347). And indeed, in these debates surrounding the touchy subject of sex trafficking of foreign women into South Korea's camptowns the particular international travel document, the E-6 visas that are given out to entertainers, came under much scrutiny. The category of E-6 visas was first created in 1993, and in 1996, the Korean Ministry of Culture and Sports officially granted the Korean Special Tourism Association (that is, a powerful organization of club owners working in camptowns) the right to invite entertainers from abroad to work in the GI clubs (Moon 2010a: 342). At first, women were recruited from several parts of the world, but eventually, club owners "found Filipinas and Russian women most suitable for their businesses" (2010a: 342). In 2001, the highest ever number of women entering the country on an E-6 visa was reported; 8,586 foreign entertainers came to South Korea with the help of this visa scheme, with 81.2 percent of those arrivals at that time being female (2010a: 342). Perhaps due to the mounting pressure to curtail the influx of women into the GI entertainment industry, however, by 2009, only approximately 2,300 Filipinas came into the country through E-6 (Rabiroff 2009). Their numbers seem to be on the rise again currently, though, with a total of 4,940 foreign workers entering the country on an E-6 visa in 2013. Approximately 70 percent of these visa holders today hail from the Philippines according to government estimates (Lee C. 2015).

Filipina migrant women, who, together with Russians, have made up the majority of the new entertainers in these GI clubs over the last couple of decades, have indeed been coming to camptown areas in relatively large numbers[10] since the mid 1990s. As we have seen in the previous chapter, by the early 1990s, when Yun Kŭm-i's murder threw a glaring spotlight onto kijich'on spaces, these neighborhoods close to US military installations had turned into economically destitute areas, where, due to a combination of factors, the number of local women working in the GI clubs had dwindled drastically. Incidentally, at around the same time, the Philippine government—another key military ally of the United States—took a historic step: the US Armed Forces, which still maintained a sizable military presence with several large installations on Filipino soil, was asked to leave the country after the base-agreement regulating the US military presence in the country was not renewed in a landmark vote of the Philippine Senate on 16 September 1991 (Simbulan 2009). Both the Subic Bay Naval Base (once the largest US naval installation overseas) and the nearby Clark Airbase were subsequently abandoned. The local economies adjacent to these bases, however, had also been heavily dependent upon the influx of young soldiers, with the adult entertainment industry, in particular, flourishing close to the bases. At the height of US military activities in the Philippines, Clark and Subic together were estimated to have hosted 55,000 prostitutes (Santos 1992: 37). Given the historical disjuncture between the United States, the Philippines and, by extension, South Korea, where troops would stay on, an informal and lucrative solution was found to the camptown crisis of the early 1990s, with the surplus of potential sex workers in the Philippines coming in handy for GI club owners located in South Korea.

Over recent years, however, the continuation of these transnational migration circuits into Korean GI clubs has proven to be a rather fragile affair. In addition to unwanted media and NGO attention, Korean club owners are today often struggling economically, as they have also come under much pressure from the US Armed Forces. Military authorities regularly put their venues "off-limits" nowadays, with any service member found in an off-limits establishment automatically in violation of US Forces Korea (USFK) regulation, which can lead to disciplinary consequences.[11] This seemingly drastic measure is one direct outcome of the "Zero-Tolerance Policy" that was launched by the US Department of Defense with regard to human trafficking in September 2003, a policy that, as Moon Seungsook critically notes, in the South Korean case mostly entailed a concern with:

> distancing the military and its soldiers from camptown prostitution tainted with trafficking. The policy is nebulous about the use of (poor) women's sexual labor presumably without force or trafficking to keep male soldiers docile and useful. It continues to normalize male soldiers' heterosexual entitlement at the expense of marginalized women. (2010a: 350)

In reaction to such concerted crackdown efforts in kijich'on, the methods deployed by club owners and recruiters to sustain the influx of young female bodies from the Philippines into the camptown areas seem actually to have become more unscrupulous. Apparently as a result of the introduction of the anti-trafficking laws of 2004, any person applying for the contentious E-6 visa ("entertainer" visa) today needs to prove that they are actually capable of singing or dancing. This is usually done by first submitting a video of one's performance skills to the recruiters, with successful applicants also being made to sing in front of Korean consulate personnel to prove that they are indeed entertainers in the narrow sense of the term (Lee C. 2015). This lengthy process may have led to the unfortunate outcome that Filipino recruiters nowadays have a harder time finding women who fit the bill, so to speak, that is, performers who can pass these tests, and who are also fully informed that they will in fact be working in the adult entertainment industry.

What is more, while these detailed bureaucratic processes may at first glance give the *impression* of legality (including to women who might be somewhat uncertain about the nature of the job they are signing up for), quite often this kind of recruitment is actually illegal under Philippine law, which some of the prospective entertainers who sign up for the job seem not to have been entirely aware of (see also the next section of this chapter). In order to recruit women in a lawful manner, Filipino recruiters would also need to acquire an Overseas Employment Certificate, which is a document issued by the Philippine Overseas Employment Administration that is only handed out after Philippine authorities have been able to verify whether the workplace abroad is safe and legitimate (Lee C. 2015). In an effort to circumvent these additional procedures that these GI clubs would most likely have difficulties passing, many recruiters these days seem to try to fly the women out of the country on tourist visas, usually avoiding the direct route from Manila to Seoul that may arouse suspicions, and hiding the women's E6-visas by gluing together two pages of their passports to prevent detection through border control personnel (Lee C. 2015).

While such techniques have for now allowed this business to continue, in spite of the increased attention paid to women recruited into the entertainment industry of South Korea, at the same time awareness of the often illicit nature of these work opportunities in Korea is slowly but surely growing in the Philippines. During a stay in Subic Bay, Philippines, in 2014,[12] for instance, I was sitting in a taxi when a local radio station was broadcasting a warning to the general public, asking women to be on the alert when it comes to job offers from South Korea that involve "entertainment." Having lived in the Subic area for a few months at that point already, I had some doubts as to whether such awareness campaigns would really be effective, though: Korean beginners classes for would-be migrants were offered in several locations in the area, and a sizable Korean expat community in Subic and elsewhere in the Philippines was certainly contributing to spreading the image of South Korea as a very prosperous nation. At the same time, the number of Filipinos living below the poverty line, according to conservative estimates by the government, stands at approximately 25 percent today (ABS-CBN News 2015),[13] with single mothers, in particular, being targeted by recruiters looking for new volunteers for East Asia's sex and entertainment industry. Incidentally, our taxi was speeding through a local red-light district, Barrio Barretto, when I heard the radio announcement in question; this little entertainment town has in the past served countless sailors coming through the Philippines and, even these days, local club owners still do not have much trouble recruiting sex workers from remote provinces of the Philippines, who will then cater to the needs of retired US military personnel and sex tourists from across the world. The business of sex in the Philippines, despite the official departure of US forces two decades ago, is alive and well, with South Koreans playing a significant role as both club owners and clients in the Philippines, too.[14]

Foreign Camptown Women and Their Management of Stigma:
"Until the Whole House is Finished"

Raquel is a 31-year-old mother of three, whom I meet at a women's shelter near P'yŏngt'aek.[15] She had come from the northern Luzon region in the Philippines to this town south of Seoul in order to work as a singer. Recruited by an agency in the Philippines that was run by a young and beautiful woman who had previously worked in a club in South Korea herself, Raquel and several other women went through a recruiting process that included putting together a portfolio of songs to perform, going through several interview stages and even a final performance in front of South Korean consulate staff. In the end, Raquel was chosen, together with two other women—Emily, who is 29 years old, and Audrey, who is 21. "I had heard some rumors about the job, read some stuff online that the job was not only about singing," Raquel told me, but she nevertheless hoped for the best before her departure. Her sister had gone to Singapore as a maid after all, and things had worked out well for her there. Her friend Emily says that her family suspected, too, that this job might entail sex work, and so some of her relatives came to the agency a few days before her departure to Korea to speak with the woman running the agency one more time. "We will take extremely good care of your daughter," the woman assured them, and so Emily finally also made up her mind to go.

Raquel, Emily, and Audrey were separated at the airport in Inch'ŏn after arriving in Korea, and Raquel was taken to Anjŏng-ri, while Emily went to Songt'an, another camptown near P'yŏngt'aek that mainly services US Air Force personnel stationed at Osan Air Base. Audrey, the most inexperienced of the women (Emily and Raquel claimed that Audrey never even had a boyfriend before), was taken to a third location, and Emily and Raquel would never meet her again. Raquel described her first evening at the club as eye-opening—in no time, she found herself on stage, performing a "sexy dance" at the pole. "Me!" she laughed, reminiscing about this evening, "a mother of three! Look at my hips!" The other women working at the club spent most of the evening filling her in about the job. Going on bar fine, they would explain to her, was the only way to really make good money around here, and they encouraged her to get over her fear of sex work as soon as possible if she wanted to do well for herself in this town: "When I think of my first guy," one of them explained to her, "now I just think of the door to the new house I will build in the Philippines. The second one, he's the window. And so it goes, until the whole house is finished."

On her second night at the club, she had her first client—an African American GI—who kept sitting with her all night buying her drinks and trying to coax her into sitting on his lap. "'Honey,' he said to me, 'tomorrow I will come back for you, and then we fuck!'" There was to be no tomorrow, though, because in nearby Songt'an, Emily had spent her first night crying in a corner of the club. She went through a lot of back and forth, Emily told me, trying to make up her mind what to do about the situation she found herself in, and the thought of her family back home in the Philippines certainly was a factor in her hesitation to call for help. "What will they think about me," she said, "getting myself into this?" Eventually, she made up her mind that she could not stay at the club and called for help via email. A Filipina friend she wrote to then passed her name and that of her club on to a friend in Korea, who then called the NGO workers from Turebang.

Within half a day, a few Turebang staff members showed up at the club, together with the police, who demanded Emily's passport back from the club owner so that Emily was free to go. She was then taken to a recently opened shelter facility run by Turebang (where I would eventually meet her). The next morning, she joined an NGO worker on a walk through Anjŏng-ri, where they were to hand out flyers to other Filipinas they met in the streets. They ran into Raquel that day, who was extremely happy to see her friend again—and they both decided to make use of the Turebang offer of a free flight ticket back home. Their agent in the Philippines, as soon as she heard about the fact that both women had left their clubs, called up Emily's parents to demand the money back that had been invested in Emily's trip overseas, but Emily's relatives threatened the agent with the police, which eventually brought an end to the stream of phone calls they received. When I last talked with the two of them, on the day before their departure, Emily said to me, "You know, I can really understand how women are drawn into this business. And why they stay, even when it gets tough. Maybe staying is easier for them than going home empty-handed, like us."[16]

The story of Raquel and Emily, who, by all definitions of the term (Schober 2007), were trafficked to South Korea under false pretenses, may still serve us as a reminder of the various complexities and social pressures that contribute to women staying in the Korean sex industry servicing US soldiers—familial obligations (that is, the pressure to send remittances home), numerous debts acquired on the way to Inch'ŏn airport, shame as well as the desire to make your fortune regardless of the odds stacked up against you are all factors that have convinced numerous Filipinas to stay at the GI clubs. If they decide to stick around and fulfill their contractual obligations they have entered into with the club owners, they soon find themselves part of quite a claustrophobic environment. The migrant women barely ever leave the camptown spaces they come to live in, as the 6 to 7 days a week they spend working at the clubs barely gives them the time and leisure to venture out into the rest of the country. In addition to the club owners and the older women managing them in the clubs, they rarely ever meet any Koreans at all. Their primary social contacts are US military personnel and other migrant workers, and their visions and hopes for a better life are often entirely focused on the GIs they get to meet during their working hours in the clubs.

Angie, a 34-year-old Filipina, is a good example of a female migrant worker who has placed all her bets on her fiancé, Bill, who is in his mid 40s. I met Angie for the first time because she had a medical emergency—a week earlier, she had had an accident and broken her arm. She had refrained from seeking medical treatment right away because she was living in Korea without documents, had no health insurance, and depended solely on Bill for money, who was strapped for cash so close to payday. "I'm sorry, I don't speak any Korean," Bill immediately apologized to the Turebang worker and the doctor on duty at the hospital where we had brought Angie when he arrived a couple of hours later. "It's just that the only people I ever get to hang out with are other Americans or Filipinas up there in Tongduch'ŏn." Bill and Angie, I learned, had met each other only a few months earlier. Because of their encounter, Angie had finally decided to run away from her club. She had continuous trouble with receiving her promised wages, suffered from the stressful working conditions at the club, and felt bad that her boyfriend had to pay large sums of money to the club owner just to see her.

The oldest sibling of eight and hailing from a peasant family, she had come to Korea from the Visayas region of the Philippines, where her two children were still living with her mother, who tried to scrape a living for the family by selling fish at the market ever since Angie's father had lost his small plot of farm land. Angie's Filipino boyfriend, the father of her children, died a while ago, leaving her as the main caregiver for her children. And even though Angie helped her mother with the selling of fish, there was never enough money. So when Angie heard of the opportunity to work as a Karaoke singer in Korea, she took the chance and applied for the job. The work turned out to be much harder than she could have imagined, though: she had been promised $800 per month, but she soon found that there was barely any singing to be done, just drinking every night with her GI clients and coaxing them into buying her more overpriced drinks which would entitle them to 20 minutes of her time.

The guys would pay $10 per drink, of which she usually received $1. Her quota was to sell 100 drinks per week, a goal she rarely ever reached, with business going very badly at times, so she typically ended up with much less money than she was supposed to make through these drinks.

Additionally, the club owner withheld her fixed pay of $100 per week, arguing that Angie would only get to see that money when her contract was finished. Under such conditions, she explained, many of the women she knew resorted to prostitution to augment their low salaries, or were talked into it by their club owners or their clients. "You know about the bar fine, right?" Bill asked me at this point of Angie's narrative, and shook his head in disgust, "Angie was so lucky that they didn't have that at her club." Angie jumped in now, "The women only do it because they are so poor, you know?" She herself never considered prostitution an option, she said, but instead she frequently ended up hungry and broke. Most of the money that she made went toward buying the sexy clothes she needed for her work, and toward buying food for herself—as per contract stipulations, the club owner was supposed to provide them with three meals per day, but instead, they frequently only received one meal.

When she met Bill, things changed for the better, though: he talked her into running away from the club, which she did after only three months of employment there. He got her an apartment and promised to marry her, as her visa status had expired the moment the club reported her as missing.[17] In the meantime, they had already signed all the necessary papers for their marriage to be legal, Bill told me, but with two different embassies and the US military involved, it took ages for all the paperwork to be processed. Bill had no intention of returning to the States any time soon, but instead wanted to continue his career with the USFK or wherever else the military would choose to send him. Angie wanted to be reunited with her children, once their marriage and her residence permit as a military dependent came through. This was bound to complicate things for Bill further, but he was confident they would be able to make their new life as a transnational family work.

Camptown Preoccupations: "Marry a Nice GI ..."

Where exactly she was living she would not be able to disclose, Angelina, a 30-year-old Filipina told me. Her fear of the immigration authorities was too great. She and her child had been living without documents in South Korea for four years now. Trying to avoid deportation, she only left her flat at night to go to work at the club. She had decided to stay in this country regardless of her legal status until her former boyfriend, a US serviceman, started to pay alimony for their child. She was currently in the process of suing him in a Korean court, but just this morning, her lawyer had told her that her case seemed rather hopeless because the young man in question had already returned to the States and refused to even acknowledge the court order from Korea.

Angelina, a very pretty woman, gave off a certain nervous, yet determined energy on the morning I spoke to her. "I only had coffee for breakfast and then we had to rush to the court hearing," she explained. She took the previous night off from working at the club in order to get all the official meetings done. She had come to Korea five years earlier, she told me, sent by a Manila-based agency. Her boyfriend back then, a European, was regularly taking drugs, so she wanted to get away from him for a while and see a bit of the world. Before that, she had already worked at a club in Japan, a totally different line of work, she first tells me—but later conceded, yeah, sure, they wanted her to sell sex there as well, and things were not going very smoothly, which is why she changed employers so many times in Japan.

The agency asked her before she headed off to Korea what kind of clubs she wanted to work at—at a place that serves GIs only, or at one where regular Asian men go as well. She picked the one reserved only for GIs, "of course," she told me, and, within three weeks, she held in her hands her visa for Korea that would allow her to work for six months. Once she arrived at her club in Korea, however, matters quickly escalated with the owner of that business over the issue of sex:

> When I got here, after that I found out what kind of job they wanted me to do. [...] They wanted us to go out with the customers, you know, "bar fine". And I told them that this was not in the papers that I signed in the Philippines. So if you want, send me back to the Philippines, I won't do this, I told them.

It is one thing, she told me, if a girl *decides* to go with a GI to make that extra money, and quite another if the club owner orders you to go with a guy: "No pushing. If some of the girls want to work that kind of job, make the extra money, then I can't stop them. But if the girls don't want to work like that—no pushing!" Within two weeks of her arrival in Korea, the police had shown up—a girl from the club had run away a little while earlier because of the conditions at the club, and a priest who had helped her escape had notified the police. The women were then all taken into custody, but jointly decided to return to the club a few days later because neither the police nor the priest could provide them with what they really wanted: a new job to make a living for themselves in Korea.

The next few months turned into a constant struggle for Angelina. In the midst of trying to negotiate with the bar owner, her GI clients, and the other girls for better working conditions, she ran away from her place of work several times. On one occasion, a GI customer hid her in his flat inside the base for an evening, but reconsidered his offer to let her stay there after the club owner threatened him with the police:

My friend, boyfriend,[18] I think, he said, just come with us on post, we can buy you everything. When we got on post, they just let us borrow their clothes and we could take a shower there. No shampoo, no nothing. Then we stayed on post until around 11. Then my club owner tried to call them, and the soldiers got scared. Because he said, he'd call the Korean police, blabla, so they got scared. So they said, "You have to run away, 'cause we're gonna get in trouble [if you stay here]." So they found an apartment for us.

She changes clubs, she changes jobs, and she changes boyfriends. She receives an offer to work "on post," that is, as a civilian working inside the base, she works the telephone there, she works behind the counter at the PX store, there's this guy who remembers her from her club days and soon enough, they start dating. He is already married to a Russian woman, he tells her that his marriage is only a sham so he can live off base. They move in together, and she soon realizes that he cheats on her, sleeps around, so she threatens to leave him, and eventually she leaves. She goes back to work at a club, where she finds out that she is pregnant. Her ex says he will pay for the abortion, but then she changes her mind, she is a Filipina after all, a proud Catholic, and once she sees the baby's heartbeat during an ultrasound, she makes up her mind to keep the child.

He refuses to talk to her now, then he says he will support her, then long silences again, the occasional phone call in between many, many unreturned ones. She is broke, too pregnant to work, friends help her out, she works in a restaurant for no pay, just for the food, enough to not starve. She runs into her ex-boyfriend occasionally, they have horrible fights and she fears she might lose the baby because of all the stress. She gives birth, he pays for the hospital bills, and he makes sure the baby gets an American passport, but then does not contact her again. She files a complaint with the military authorities, and his supervisor apparently speaks to him, and nothing comes out of that. He doesn't pay for any of her additional costs, he is dating another Filipina now, he talks badly about her in front of her friends, he keeps telling her that she should give him the baby so he can take it to the States. He wants to marry the other Filipina now, who has a kid of her own, from yet another GI, they want to raise both children together in the States. Angelina goes into hiding, because she believes he wants to call immigration on her in an attempt to get the child. She changes flats, she changes jobs, she is in a constant panic, until the day that he leaves for the States.

Would she like to go back to the Philippines some day, I ask her. Maybe later, she says, but she wants this child support issue solved first, she wants a future for her boy—she cannot give up, not just yet, and how are they going to live in the Philippines, she asks me. Her little boy is an American citizen, so they will have to regularly leave the country, "With what money, how are we supposed to do that?" The last time she spoke to the father of her child was six months ago. Her son was very ill at that time. "If you had given me the baby, that kind of stuff would not have happened," he told her on the phone. His Filipina wife later wrote her an email, starting it with "How's night life in Korea?"—"She speaks as if she's never seen a club from inside herself," Angelina laughs bitterly. What about the future, I ask her. "Marry a nice GI," she says, wait and see, and then pick the right one, make sure she can go to the States with him and file a law suit there. "Have you met anyone nice yet?" I ask. "No, there's no one," she says, and there's tears in her eyes now, it has been a very long day for her.

Two hours later, we sit on the bus that is bound to take us back to the town she lives in. Her son is excited to be on the bus, he puts his little fingers onto the window screen—as we depart, we pass by Camp Stanley, and its many buildings, vehicles, military structures come clearly into view for a brief moment. "Mama, look!" the child exclaims, pointing at something he sees inside the base, and so they both look, as the US military base and all that is inside of it rapidly disappears out of sight again.

Villes as Captured Spaces

When camptown spaces were first cast into view as violent spaces from 1992 onwards in Korean public debates, they proved to be extraordinarily fruitful zones of the imagination that gave nationalist actors a powerful tool that allowed for a drastic repositioning of South Korea in relation to its powerful ally, the United States. Whereas the public focus on kijich'on is primarily directed at how these realms are endangering to local (and, to a lesser degree, to foreign) women, in this chapter I have shown that a crucial difference between discourse and practice, between imagination and lived experience, can be made out when looking closely at the ville. Camptowns, as they present themselves in their 21st-century reconfigured forms, are much more transnational spaces for asymmetrical encounters between GIs and local and foreign women employed in the entertainment areas than they are spaces of outright domination that pertain to a nation's fate. No doubt violent escalations are, to some degree, normalized experiences in people's lives in these areas, but the story cannot, and should not, end on this note. Murder and rape are not daily events in kijich'on for the women employed there, but fearful suspension in between different countries and legal systems is, made worse by financial and emotional difficulties, and by finding oneself in a dire spot far away from home.

Camptowns are both endangering and endangered spaces that require much maneuvering, scheming, and everyday strategizing by those who live, work, and play in them. Alliances shaped in kijich'on—of a merely sexual or romantic kind—are made on a daily basis, but in this transient space, seemingly stable commitments often collapse with great ease. Hopes for a better future always threaten to evaporate into thin air, and the soldiers' money seems to be constantly slipping through the fingers of those who seek to get their share of it. And in the midst of all the young bodies encountering each other in the dark spaces of the GI clubs, the older ones, cleaning the counter, filling the glasses anew, or mopping the floor, may easily be overlooked—but it is these older Korean women, having experienced double abandonment by their co-ethnics and their former American lovers, who are often the only ones among all of kijich'on's actors, who are here to stay.

To be sure, a few of their friends have done well for themselves by building up new lives with money earned in the ville or by founding transnational families with their soldier lovers. It is with similar success stories in mind that many of the Filipina women go about their business in kijich'on—with their lives seemingly on hold, suspended between state of failure and success, hope and abandonment. Camptown is an *affective* space that they find themselves temporarily caught up in, a space that, in the way that it also holds the potential to throw them onto new life trajectories, is quite *effective*, too. Camptowns, it is crucial to note in this context, are not only spaces that capture those foreign men and women for a limited time, they are also captured spaces, semi-occupied terrains that, due to the proximity of the base, are seemingly not quite part of South Korea any more. The women's actual "labor of love" (Yea 2005), I have argued, can also be read in light of this military appropriation of these spaces—the entertainers' intense preoccupation with their soldier-clients,

I believe, cannot easily be separated from the fact that these encounters occur in the midst of a heavily militarized environment.

Camptowns are usually doubly removed from the center of gravity in the region, that is, downtown Seoul, with the introduction of foreign sex(ualized) labor to serve the proletarian military labor of the United States further disconnecting these areas from Korea proper. For many of its inhabitants a kind of suburban marginalization—and the social and geographical immobility that comes with it—is a very real fact of life indeed. While the older women often find that they have nowhere else to go, the younger ones see that their mobility (both across Korea, the region and the globe) is very much dependent on how well they play their cards at the game of the night in the GI clubs.

US soldiers, on the other hand, can and do move relatively easily across the greater landscape that is the Seoul Capital Area during their time off from work. In addition to the troop reductions and relocations that have had great impacts on camptowns in the past, nowadays another more silent threat to the ever-dwindling prosperity of the villes can be found in easy access to public transportation, and in the presence of glitzier entertainment districts a mere hour or two away, where sex can often be attained without monetary compensation, and possibly also without all the intense emotional labor that preoccupies both entertainers and soldiers in the camptowns. Faced with the progressive devaluation of kijich'on spaces, the soldiers—as will be explored in the next two chapters on It'aewŏn and Hongdae—increasingly deploy a strategy to get themselves out of the villes. In their free time, they often seek out inner-city entertainment districts instead, with their claims to a right to downtown entertainment dis-placing and dispersing much of the old conflict over their presence in the country across the vast urban space of the capital. At the same time, the women brought in for their entertainment once again find themselves left behind.

5

It'aewŏn's Suspense

Of American Dreams, Violent Nightmares, and Guilty Pleasures in the City

Militarized Masculinities at Play

"Welcome to It'aewŏn. Welcome to Asia's paradise," the old South Korean man selling paintings next to Burger King says to me in English the moment I exit from the subway. "Welcome to my playground," says Joo-hwang, a 23-year-old Korean student who is waiting for me there, as he steps closer. While we walk along the crowded main road toward a club, squeezing our bodies past food stands selling grilled chicken, dumplings and kebab, Joo-hwang—who is already a little tipsy—shows me his arm. "Look, I've got plenty of tattoos tonight, don't I?" I glance at Joo-hwang who shows me all the stamps on his skin from the various venues he went to earlier. A US soldier, incidentally crossing our path, overhears Joo-hwang's words, and starts to laugh now, winking at me: "I could show you a few tattoos, too. Real ones, baby. Wanna see?"

It'aewŏn is a Seoul neighborhood within walking distance of the US Army garrison Yongsan—a military installation that at the time of my fieldwork was home to nearly 17,000 service members, civilian US state employees and their respective dependents (Powers n.d.). Consequently, It'aewŏn is the entertainment area in Seoul that is most heavily frequented by US soldiers in their "down time." During the brief moment described above, which played itself out in this neighborhood between a Korean student, a US soldier, and a European anthropologist, tattoos—and their significance as emblems of hyper-masculinity—came to the forefront in a little mock-fight over my attention. While many US soldiers I met in the country were sporting quite a few tattoos, only on rare occasions would I encounter Koreans who had undergone this bodily modification

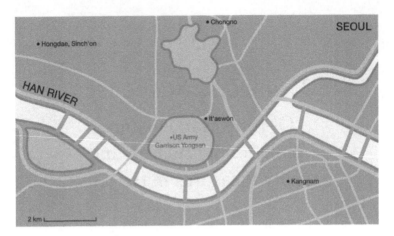

Figure 5.1   Map of inner-city Seoul

as well. Joo-hwang, too, would never consider getting one, and his goofing around with the stamps on his hand very much fits the role he wants to play whenever he is out in It'aewŏn. The neighborhood, to him (and to many other visitors of this area, I was to find out), functions as a quasi-carnivalesque territory of make-believe that lies outside of the social, geographical, and temporal parameters of "Korea proper." Being in this particular place, and in the company of a foreign woman, Joo-hwang went along with the moment. In light of this, the random encounter with a US soldier proved particularly irritating to him, because this stranger in one sentence called him a fake by pointing out that he, the soldier, was the proud owner of the "real thing" that Joo-hwang was just pretending to possess, and that Joo-hwang had perhaps set a foot too far into GI turf.

It was not the first awkward moment with a US soldier in his life, Joo-hwang later told me over dinner. For the previous two years, he had in fact been a KATUSA,[1] serving inside a US Army unit at the nearby Yongsan garrison. "I got lucky," he said to me about how he had been chosen as one of the few to serve his mandatory military time inside the US Forces Korea (USFK) headquarters. In this way, he could improve his English skills—an invaluable resource in today's highly competitive Korean job market—and also use the manifold recreational facilities that Yongsan has to offer and that, in his view, are so much better than the ones Korean military installations have for their own soldiers. His time spent at Yongsan, however, left him deeply ambivalent about the Americans he worked with: with some of the soldiers, he formed friendships while drinking and partying after work in It'aewŏn. Others, he found, would only treat him with contempt. He brings it down to a masculinity issue when explaining the recurrent antipathy to me that marked some of his relationships with US servicemen: "They just don't think we are real men like they are." Joo-hwang strived for the "flower boy look"[2] so popular in Seoul when I first met him in early 2009; he visibly spent much of his money on good haircuts, nice clothes, and accessories. Today, he has also brought along an expensive-looking man-purse, an item that has become increasingly popular with fashion-conscious young men in Seoul. Pointing to his bag, he laughs now: "You see, *this* kind of thing GIs most certainly don't understand."

As laid out in chapter 1, in much writing on militarism and gender that focuses on the US military, US imperialism is imagined as an enterprise that is significantly held in place by a kind of virulent masculinity performed by soldiers in the entertainment areas near US military bases, where they are understood to seek domination over the local population via the bodies of women. And the entertainment area that US troops in Seoul primarily inhabit in their free time has certainly been turned into an area that functions as a breeding ground for the creation and maintenance of hyper-masculine identities. However, such a focus on US bases overseas as *only* generating destructive gender relations imbued with violence among its allied nations may inadvertently conceal some rather remarkable unintended consequences of the prolonged and uneven encounter between US military and Korean civilians. And It'aewŏn, in particular, certainly holds a plethora of surprises.

In this chapter, I shall explore two additional factors in the complex equation of (in-)security created by US forces in South Korea: first, local men and their at times rather perplexing entanglements with the foreign soldiers on their territory need to be actively brought back into the analysis. Joo-hwang's unresolved tensions toward the Americans he considered his comrades for the duration of two years is in a way symptomatic of how the "tensions of empire" (Cooper and Stoler 1997) may express themselves at the level of young local males. If we follow Catherine Lutz's highly useful approach that "imperialism [can] be seen as a complex of concrete social relationships entailing culturally constructed emotions, ambivalences, and ambiguities" (2006: 595), then much can be gained if we explore the multiple performances of masculinity this gives rise to in the gray area between the USA and Korea.

Second, I shall further look into how the spatial dimension of US militarism overseas is constituted and contested in inner-city Seoul—that is, how the actual contact zones in which US soldiers and South Korean civilians meet are (re-)configured in the highly urbanized core of the larger capital area, rather than in the suburban outskirts that we have explored in chapter 3 and 4. Within debates on globalization, a focus on the production of specific places through the interplay between larger, outside forces and local agents was proposed as a solution to the conundrum of how to conceptually link up the global and the local in a meaningful way (see, for instance, Appadurai 1996; Burawoy et al. 2000; Gupta and Ferguson 1997; Tsing 2000). Place, in such an analysis, "becomes a launching pad outward into networks, backward into history and ultimately into the politics of place itself" (Gille and Ó Riain 2002: 287). Informed by an understanding that the ethnographer is "less a chronicler of self-evident places than an interrogator of a variety of place-making projects" (2002: 287), we can better ascertain how the mightiest armed forces in the world come to "take place" in this community, which historical contingencies have shaped this special region, and what the stakes are for civilian actors who seek to shape this territory for their own benefit.

After giving a sense of the compressed heterogeneity that makes up the neighborhood, in the following section I shall use the launching pad of It'aewŏn to briefly hurl us back in time once more in order to touch on

Figure 5.2   A side street of It'aewŏn (nearby Homo Hill)

some of the social, historical, and geopolitical parameters that gave rise to the district.[3] I will argue, in particular, that a containment strategy of sorts was at work during the Cold War period with the aim of keeping US influences within the boundaries of the district. Second, the arrival of sexual and ethnic minorities in the neighborhood since the beginning of democratization will be contextualized, with the multifarious ethnic, religious, racial, and sexual identities on display these days increasingly being turned into commodities fit for consumption. Following this, I will look at a distinct ambiguity that presently characterizes the area: the uneasy positioning of the neighborhood between allure and repulsion that seems to dominate many people's imaginaries, and that is a phenomenon I call It'aewŏn suspense. The at times very rowdy practices taking place in this neighborhood prove to be both dangerous and creative, and have simultaneously engendered destruction and production of social meaning and order. It'aewŏn's suspension between competing sovereignties that express themselves in highly gendered ways on the ground, we shall see, has provided some groups of people with rather counter-intuitive liberties in the shadow of the US base.

"Special District It'aewŏn": Of Containment and Fermentation

If a visitor, pushed along by nightly crowds, were to make her way up the main road of It'aewŏn, she would quickly leave behind the many clothes stores, souvenir shops, international restaurants, and Western-style coffee shops that more family-oriented tourists like to visit during the daytime, to then approach the most contested and infamous corners of the neighborhood. Next to It'aewŏn's fire station, a smaller street leads into a hilly area with a diverse range of bars and clubs crammed into a number of older buildings, some of which have young Filipina or Russian women loitering at their entrances. A few steps further, she would be faced with a dark alley cutting across the street that is dimly lit by the fluorescent lights of a succession of small venues with hyperbolic names ("Best Club," "Tiger Tavern," "Texas Club")—this is "Hooker Hill." Past the signs in Korean that warn minors to stay out of this area, you can see scantily dressed Korean women casually standing in the street who try to convince the few lonesome male figures making their way up here to set foot in their premises.

A few steps further down the actual street, running parallel to "Hooker Hill," one may find yet another small alley that is commonly known by another peculiar name: "Homo Hill." While "Hooker Hill" is usually eerily quiet, "Homo Hill" is on your average Saturday night filled with the laughter of many young men who come here to drink, dance, play, and make out in front of gay and transgender clubs and bars. Drag queens in bunny costumes hang out on the side of the street, chat in fluent English with visitors and have their pictures taken with whoever comes their way. As in most other bars of It'aewŏn, the waiters here, too, speak both English and Korean perfectly well, US dollars are accepted as a second currency, and locals mingle freely with foreigners, with women being a curious oddity in this part of the neighborhood.

Just a two-minute stroll away from this lively part of town, It'aewŏn's Catholic church is located. And several moments later, one will come across the so-called "Muslim Street" that runs parallel to "Homo Hill" and features Halal shops, Western Unions, various Middle Eastern restaurants, and clothes stores that sell the hijab. It leads toward South Korea's biggest mosque, the Seoul Central Masjid, which is situated at the very top of this hilly area, overlooking the contested space underneath it. Business on "Muslim Street" is booming, particularly on Fridays and the weekends, as most of the predominantly male Muslim migrants frequenting the area do not live around here, but merely come to shop, visit the mosque and socialize in the neighborhood during their time off from work. In this way, racy sex establishments can be found practically side by side with various religious institutions in the narrow space that makes up It'aewŏn.

It'aewŏn's intensely foreign quality has a long tradition, and so does its association with war, colonialism, and foreign exploitation. One rather mythical-sounding story is that during the 16th century, when Japanese troops first invaded the peninsula, they came upon a temple in the area that is today's It'aewŏn, where they raped the local nuns and impregnated many of them. After the foreign invaders left Korea, the women were made to give birth and raise their children in the vicinity of the temple, with the Korean authorities also sending surrendering soldiers to this area. In accordance with this narrative, then, some have argued that the Hanja behind "i-t'ae-wŏn" means "people from a foreign land" (while others claim it simply refers to this region as once having been rich in pear trees) (*Stars and Stripes* 2009).

Regardless of whether the story about the nuns is founded on fact or fiction, some features of the narrative—hinting at a strategy of strict territorial isolation of foreign influences in one region—most certainly would be in accordance with the hermit kingdom politics pursued by authorities in the Chosŏn dynasty (1392–1910) (Cumings 1997: 87ff), when this occurrence supposedly took place. This strategy, as we have also touched upon in chapter 2, would fail terribly in the late 19th/early 20th century, when Korea's closed door was forced open by neighboring Japan, which sought to get ahead of Western powers in the scramble for territory in North East Asia. However, the notion that the enticements of the foreign—and the bodies that are seen as capable of spreading these wherever they walk—were best contained in one territory found another crucial defender in military dictator Park Chung-hee (1917–79).

In 1904, on the brink of the country's forced inclusion into global modernity through Japanese colonialism, It'aewŏn became a military stronghold for the occupying forces. The actual city of Seoul was then still miles away from the area that is today's It'aewŏn, and the Japanese continued to occupy the area until 1945, when the US 24th Army corps took over the Japanese headquarters left behind by the retreating troops that had just lost the Asia-Pacific War (1941–45). Following the country's division by Soviet and US powers in the same year, and the Korean War (1950–53) that was to arise out of it, American forces held this 2.5 km^2 area until today, a very visible spatial marker of South Korea's complicated and highly uneven relationship with the USA.

It'aewŏn, in its first humble post-war manifestations was a rather typical military camptown similar to others near US installations across Asia at that time: ramshackle houses were hastily built in order to cater to the Americans' needs, and many impoverished locals arrived seeking to make money out of the foreign soldiers. Prostitution was the most infamous of the services offered and, even though sex for sale would be outlawed by the temporary US military government (1945–48) in 1947, at the same time an "unofficial but consistent policy" was born that gave rise to a simultaneous criminalization and regulation of the sex workers (Moon 2010c: 46). A complete eradication of the sex industry next to the US bases was clearly in the interest of neither the USA nor of the Korean authorities, with economic calculations playing an important role on the South Korean part: Park Chung-hee, as we explored in chapter 2, was looking for quick ways to amass the foreign currency so urgently needed to fuel the devastated economy of his country (Lie 1998: 43ff), and the women hustling Americans in neighborhoods such as It'aewŏn proved a goldmine in this respect.

While economic benefits were perhaps the largest factor behind the state endorsement of camptown spaces during the Cold War era, their existence also allowed for the containment of US soldiers to clearly delineated areas (Moon 1997: 39). In a contradictory fashion, while the sex industry servicing GIs was being protected, at the same time Park Chung-hee was highly concerned with the morale of the local population. At times, Park saw the country plagued by a "monstrous, chronic disease," triggered by the introduction of "American things, Western things, (and) Japanese things," which in his view were to blame for the corrosion of the national spirit (quoted in Walhain 2007: 87). A localized containment doctrine of sorts that was aimed at curbing US influences (a rather surprising side effect of the US policy directed at restraining the spread of communism in South East Asia) seems to have been the preferred remedy.

The strategy proved effective at first: sociologist Lee Na Young, for instance, asserts that "the Korean government has successfully ghettoised the [camptown areas] as buffer zones to prevent US soldiers from entering Korean society and prohibit ordinary Koreans, especially 'respectable' Korean women, from interacting with US men" (2007: 454). And, with viable alternatives to a life outside of prostitution blocked for most of the women working in camptowns, It'aewŏn was seen by many as a particularly attractive springboard into a better future in the USA, or, at the very least, as a gate to a "Western" kind of life within Korea. Ms. Kim, an activist from the sex worker drop-in center Sarangbang, points to one major pull factor drawing women into the neighborhood: "In It'aewŏn, the older sisters,[4] their agency, their desire, I think, it's the American Dream. It's an important word in It'aewŏn, the American Dream. It still is today." "The American Dream" here not only stands for the widespread goal among the women to marry a US citizen, but also for the practice of conspicuous consumption, which may involve spending all your earnings in one night (many of the Korean women employed in the neighborhood nowadays came to the area in the first place because they enjoyed partying here; only later, I learned, did they turn to sex work to support their lifestyle).

Another group that flocked to the neighborhood after the Korean War was Korean musicians interested in Western pop and rock music, who were hired by a variety of venues to entertain the troops (Kim P. and Shin 2010; see also Russell 2008: 143). Following on the heels of those adventurous spirits, ordinary Korean college students and other young people started to make excursions to It'aewŏn as well, as they wanted to experience the new kind of music to be heard in the clubs (Kim E. 2004: 45f). Something was clearly brewing within the space of the illicit zone: the district to some young people not only became a window to the Western world that allowed them to acquire new tastes. To some degree, It'aewŏn—and the subcultures one could encounter in it—also enabled the expression of a certain opposition to the rampant conservatism that arose during the military regime that was to rule the country between 1961 and the late 1980s.

Fearing that this particular "taste of freedom" would spread throughout the city, in the summer of 1970, the youth that had to some degree emerged from It'aewŏn became open targets of the police. Men were given haircuts on the spot if their hair was deemed too long (Lankov 2007: 326), and women were reprimanded for the length of their skirts: "The streets of Seoul turned into a theater of the absurd, where police officers, armed with measuring sticks, imposed the 'discipline of the body' on the hapless passersby" (Kim P. and Shin 2010: 216). Because of this cultural purge, local musicians and their fans in Seoul retreated back into It'aewŏn, to wholeheartedly embrace the hedonistic lifestyle there, consuming drugs for the first time, engaging in sexual experiments, and staying up all night, thereby deliberately breaking the night curfew imposed on all Korean citizens at that time (Kim P. and Shin 2010: 219f). The notion that It'aewŏn is a container of foreign influence, together with the idea that It'aewŏn air might be a little lighter to breathe in, were thus established as conjoined during the harsh years of the military dictatorship. The very idea of territorial containment that became so crucial to South Korean leaders during these early days of It'aewŏn indeed backfired rather strongly in that it allowed for an intense fermentation and eventual leakage of those cultural influences that were to be held in place.

Liberalizing It'aewŏn: A Street of One's Own?

Jin-su is a 28-year-old Korean man, who works as a waiter in an It'aewŏn gay bar and lives in a small flat a few streets away from his place of work. He shares his apartment with three transwomen sex workers, who all work in a club just across the street from his gay bar. His flat mates are a rather rowdy crowd, Jin-su's friend Bong-hee, a company worker, tells me. "Frankly, I thought so badly about these girls before I met them," Bong-hee admits, but says he has changed his mind about them now: "It's just a job. And in fact it's the only job they can get in Korea anyways." Their way of speaking, he adds, is too much too handle at times, though: "They love to talk about sex, and sex only. And my God, do they swear!" Once, they literally made a gay American friend of Jin-su's run away—they employed their "It'aewŏn English," a mix of broken English mixed with Korean words, to interrogate his friend in the toughest fashion about his sexual preferences. "It's a true culture shock for me sometimes," Bong-hee says, "the office on weekdays, and this madness on the weekend."

Both Bong-hee and Jin-su are very carefully managing their lives outside It'aewŏn to keep their homosexuality a secret from their families, straight friends, co-workers, and acquaintances. Jin-su's parents, for instance, think he is currently spending a year abroad. They live in a more remote suburb of Seoul and would never come to this part of town, so Jin-su is certain he will not run into them by accident during his time off. His parents kept pestering him about the fact that was turning 30 soon, as they wanted him to marry a nice Korean woman by that age. In an attempt to stall their marriage plans for him, he came up with his fake trip abroad. Coming out to his family is not an option, Jin-su says, so he chooses a strategy that is quite common among the young Korean men I meet in It'aewŏn: he compartmentalizes his life, and hopes that people who think of him as straight will never stumble upon him in It'aewŏn.

The highly charged rendezvous between the US military and the civilian urban area of It'aewŏn, as we have seen, has given rise to the neighborhood as a de facto buffer zone that was to absorb potentially dangerous foreign influences, but which in the end functioned more as a fermenting container of unwanted cultural forces. With the idea spreading that the district was somehow "freer" and more liberal than other areas of Seoul, It'aewŏn would attract a new cluster of people who found themselves hard-pressed for space elsewhere: gay men and transgender people[5] have opened several clubs in the area since the mid 1990s, with the number of queer establishments now hovering around 20.

"Homo Hill" is a spatial marker of the social change that has occurred for homosexuals amid the sweeping changes that democratic liberalization brought. With Confucianism constituting "heterosexuality as a key social and ethical norm in Korea" (Kim Y. and Hahn 2006: 60), for a long time many Koreans viewed homosexuality as solely a foreign disease that was recently imported into the country, thereby ignoring historical evidence that homosexuality and transvestism has played an (albeit marginal) role in Korea's social life for many centuries (Kim Y. and Hahn 2006). Since the 1990s, however, a small number of openly homosexual student organizations have made the first few steps toward a cautious introduction of the topic to a Korean audience (Seo 2001: 71). In 2000, the public declaration of prominent actor Hong Suk-chon would first introduce the subject to a broader public. Hong had met his first boyfriend, a Dutch man, in an It'aewŏn bar, and had subsequently spent years living in Europe, where he encountered the queer movement. His later coming out in Korea certainly created the first few broader debates on homosexuality in Korea, but also nearly ended his acting career for good. Faced with a dearth of job offers, Hong eventually opened a few venues of his own in It'aewŏn (Onishi 2003).

"The reason why It'aewŏn could become such an open gay town," Jeon Bongho writes, "depends mostly on the geographic characteristic of the location. [...] In many Korean minds this place always had the exotic flair of pseudo-overseas, which is probably the major reason for the openness with which the gay community could establish itself in the neighborhood" (2005). It'aewŏn's "Homo Hill"—the quintessential space for "deviant consumption" (Kim J. 2007: 629)—is also frequented by members of the US military, who make up the second largest group of foreign visitors to the area after English teachers. Due to the (only recently repealed) ban on homosexuals in the US Armed Forces, their visits to "Homo Hill" often have a rather cautious air. With the clubs and bars frequented by their straight comrades literally just around the corner, many servicemen still seek out the opportunity to party on "Homo Hill." "This is the perfect place, if you've never been (openly) gay before, to come out," an army medic stationed at the garrison told *Stars and Stripes* in an interview: "You're in a foreign country. When you go back to the States, who knows?" he added (Rowland 2010).

In addition to the gay scene, another group of late arrivals to the neighborhood has made a great impact over the last two decades: attracted by South Korea's recent economic prosperity, migrant workers from other parts of Asia, Latin America, and Africa have started to come to South Korea (Park W. 2002). Muslims, in particular, have made It'aewŏn their second home: in 1976, the Seoul Central Masjid was erected in the area (Lankov 2007: 265ff), making It'aewŏn a natural center of gravity for many South Asian migrant workers who primarily do 3D (dirty, difficult, and dangerous) jobs in the country.

Furthermore, the number of Africans has sharply increased over the last few years. In 2004, only 385 Africans were registered as residents in It'aewŏn, while by 2009 their number had reached 706. They comprised 16 percent of the 2,388 foreigners[6] residing in the It'aewŏn area at that time, with Nigerians being the most numerous nationality, followed by migrants from Ghana and Egypt, most of whom are young, single and male (Han 2003: 163, 166). It'aewŏn became a focal point for their social and commercial activity, possibly because of the presence of larger numbers of African American soldiers in the area—the newcomers may have expected a friendlier atmosphere on their arrival in a neighborhood of Seoul that already hosted black people, anthropologist Han Geon-soo suggests. A Nigerian migrant told him that when he first arrived in the country, he got into a taxi and asked the driver to take him to an area where people looked similar to him—and "when the taxi arrived at some place, he saw many black people hanging around in the street. He was glad to see African brothers and got out of the taxi, but then realised that they were black American soldiers" (Han 2003: 166).

The recent influx of migrants can be felt in the urban landscape of It'aewŏn through the ever-growing number of ethnic restaurants, a fact that is utilized by the Korea Tourism Organization to promote the neighborhood in the following way: "Unique flavors, exotic interiors, and diverse nationalities help to make It'aewŏn befittingly 'the global village of Seoul'" (Korea Tourism Association n.d.). This kind of language is symptomatic of a wider trend. In the Seoul of today, the multifarious ethnic, religious, racial, and sexual identities on display in this entertainment district are increasingly turned into commodities fit for consumption. In this way, the democratic and social liberalizations, which have partly emerged from and partly acted upon the place of It'aewŏn, have ironically also opened the gates for rampant economic liberalization within the neighborhood.

The widely publicized plans to turn the Yongsan garrison—2.5 km^2 of highly valued property in the center of Seoul—into a gigantic public park once the US Armed Forces hand the area back to the South Korean authorities have certainly been slow to be realized. First announced in 2004, the plans to relocate most of the troops stationed in Yongsan to P'yŏngt'aek and Taegu have been delayed repeatedly, with the US authorities nowadays negotiating to retain a small military presence of a few thousand soldiers in Seoul's center after all (Park B. 2014). These relocation plans have still managed to further invigorate speculation on the future rise of It'aewŏn's market value. Within this process, some of It'aewŏn's more vulnerable inhabitants are left behind, desperately scrambling to keep their hold on what is essentially their neighborhood and home. The Korean sex workers running the small businesses on "Hooker Hill," many of whom have also been living in the area for years or even decades, increasingly find the de facto protection that the vicinity of the base has given them from prosecution crumbling. "Poor people don't have a history,"[7] an activist from Sarangbang sums up their dire situation, "and when redevelopment starts they are the first who will be made to leave." With such drastic changes on the horizon, the highly regulated zone—and its dangers, which have troubled authorities for decades—is on the way toward being re-fashioned into an up-scale adult playground, where the global can be tasted and experimented with for the length of a night. As we shall see, contradictory sentiments toward this neighborhood still loom large, with violent imaginaries about It'aewŏn being part and parcel of a particular suspense that—to some—makes up much of the allure of the area.

Figure 5.3   "Hooker Hill" in It'aewŏn

It'aewŏn Suspense: Space of Pleasure, Realm of Fear

Across from F., a bar on Hooker Hill, former KATUSA Joo-hwang and I watch a white man in his 50s, who is just entering one of the little brothels lined up on this street, his big hands placed on the shoulders of a Korean woman half his age. Joo-hwang pauses for a second to watch the two of them disappear, then he shrugs his shoulders ever so slightly, and we move on. F. is relatively deserted tonight, a few young men and a handful of women in mini-skirts in between; everyone seems excessively bored, except for the people seated at the table across from us. One of the GIs sitting there is busy—carefully placing shot glasses in the cleavage of the blonde woman he is with. Then he rubs his face into her breasts, grabbing the rim of the glass with his teeth, and down goes another shot of tequila that way. Jason, Joo-hwang's Nigerian friend, arrives now, and urges us to move on to another place, Club C., a venue that is located a few steps away.

At C., a Korean DJ plays one hip hop track after another, with an MC casually picking up the mike to rap along to the recorded tunes. One Korean girl, in the shortest dress, keeps dancing on and on all by herself, and there is a group of young, blonde Russian women sitting next to their big, bald-headed boyfriends, overlooking the dancefloor with great disdain. Jason buys us a round—he tells us that he has already gambled away over a thousand dollars tonight while partying at the nearby casino inside the US base. What does a little more money spent on booze matter then, he gleefully declares, and we raise our glasses to that. Joo-hwang looks a little miserable while Jason keeps bragging about the good times he's had at the American casino. Koreans are prohibited by law from gambling, but Joo-hwang still tried to sneak into the casino the other day, with security eventually turning him away. Joo-hwang did not tell me much about his friend Jason beforehand, so I attempt some probing now. Other than partying in It'aewŏn, Jason tells me, he does not really do anything special here in Korea. Skillfully warding off my questions as to what brought him to the country, he says to me: "I'm here to have a good time, dear. And what about you?" This is indeed a crucial lesson to be learned in It'aewŏn—questions regarding people's professions or life circumstances will usually be left unanswered. The only reply you may ever receive from foreigners in It'aewŏn is "I'm in business," and that will have to do.

Half an hour later, we find ourselves in Club U., where several Filipina and Russian entertainers wait around for male clients to buy them overpriced drinks. On the dance floor, I see a few "K-Lo"s (slang for Korean women who emulate an African-American style), otherwise the place is filled with young men, most of them black and Latino GIs, or Africans like Jason. The security guys lurking at the entrance wear bullet-proof vests and carry guns, as Club U. is infamous for its bloody brawls. And, sure enough, after just 20 minutes which we spend on the dance floor all hell breaks loose: a fist fight starts among several of the dancers, and for a moment the music is turned off. The girls start shrieking, while nearly every man within reach seems to attempt to join the tussle. The bouncers get to the dance floor with dazzling speed, though, and in no time the prime offender, a big guy who puts up a fight until the very last moment, is being dragged out of the venue. Jason and Joo-hwang ready themselves to leave now, too. "Shit, all this fighting really spoils my mood, every time it's the same in this fucking place," Jason complains, and insists that we move on now; and so the night continues.

The It'aewŏn of today, as we have seen, is no longer simply a military camptown or even an urban red-light district. Among the extremely diverse, predominantly male crowds populating the streets, territorial struggles over the limited physical space of the neighborhood, and contestations over potential sexual partners to be met within the area, are the order of the day. Tensions are further heightened by the ubiquity of weapons in the district: both Korean police and US military police (MP) have a heavy presence in the area—a very tangible everyday reminder of how two national sovereignties rub against each other in this terrain, creating ever shifting layers of friction (see also Tsing 2005) on the ground that visitors of the area need to come to terms with. Additionally, most of the clubs employ their own private security personnel as well, who often come equipped with all kinds of weaponry to guard themselves against potential offenders—a kind of neoliberal outsourcing of the state task of providing security that also points to some of the recent developments of commodification and gentrification that I have briefly touched on earlier.

On one of my first few nights out in It'aewŏn, I was very surprised to come face to face with a pair of US MPs in full uniform and armed with machine guns entering the bar I was sitting in. While the MPs patrolled through this particular place—a "Western-style" bar popular with GIs located on the main street of It'aewŏn—I was in the midst of talking to my friend Bong-hee, a 25-year-old office worker who, as noted earlier, spent most of his weekends partying on "Homo Hill." Half an hour later, once we had settled down in his favorite gay club instead, I asked him whether I should expect to see MPs here, too, but Bong-hee just laughed and said that they rarely ever ventured into this part of the neighborhood: "Imagine that, these guys in uniforms with guns storming into this place. Yeah, I can picture it, 'hands up, all you gays!'"

"A lot of old people are afraid of It'aewŏn, just so you know," he had warned me earlier that day before we got on the subway. I said, "I've met plenty of young people here in Seoul, too, who wouldn't wanna set a foot into this neighborhood." He answered: "I had a friend in high school who told me, 'In It'aewŏn, the gays fuck in the street.' Naturally, I had to go check it out!" Bong-hee's statement hints at a rather important contradiction that I came across repeatedly in people's thinking about It'aewŏn. With the delineation of It'aewŏn as a geographical space that contains particular social evils that cannot be found in any other part of the city, another seemingly counter-intuitive logic arises that I want to call *It'aewŏn suspense*: in an area of such ill fame, many practices can be tolerated that do not find space elsewhere, and an even wider spectrum of people are attracted to the area in the hope of finding room for themselves. This uneasy positioning of It'aewŏn between allure and repulsion that seems to overshadow many people's imaginaries about this place is in fact a crucial affective dimension that arises out of the neighborhood's *suspension* between the various competing sovereignties that seek to control what happens in this terrain.

In spite of their many powers, which should allow the US military and Korean state authorities to engrave their vision of order into this place, matters tend to get very muddled on a daily basis. Even though the regulatory forces of MPs, Korean police, and private security personnel are indeed highly visible during an average It'aewŏn night, the lines of jurisdiction—who is to exert power over which of the district's many visitors—are often so unclear that the use of force to intervene in situations only occurs in the event of a real escalation. Also, some of the neighborhood's visitors know exactly how to play their cards right with the authorities—GIs, on occasion, seek to pass as civilians, while Nigerian migrants, if the situation calls for it, often do not hesitate to pose as black servicemen. In short, It'aewŏn leaves much room for visitors to navigate more or less successfully through this highly loaded space that is freighted with competing meanings and presents different actors with ever shifting internal and often unspoken rules.

An understanding of It'aewŏn's territory as a set of practices (as, for instance, suggested by Andrea Brighenti [2010] in his article "On territorology") might be useful here: "Connecting past knowledge to present circumstances, a practice enables us to encode and decode signs, to share a meaningful environment, or in other words, to territorialize environments" (2010: 62). With a particular territory generated by a collective "act of imagination, a prolongation of the material into the immaterial" (2010: 68), as Brighenti argues, places may be understood as physical realms which are inscribed with potentially incommensurate visions, dreams, and desires by various actors.

Bong-hee's qualms over GIs entering "his" territory of "Homo Hill" can serve as a good example of how vastly different players have to arrange themselves in the limited space available. Bong-hee's personal understanding of It'aewŏn, it is important to know, is vitally linked to what his ordinary life outside of the neighborhood *cannot* offer him. The overwhelming majority of his friends, family, and co-workers do not know about his sexual orientation, and he fears for his job, for his friendships, and the close relationship with his parents, if news about his homosexuality were to ever come out. Thus, It'aewŏn, for him, is not only a space that allows for erotic encounters with other men but also, more importantly, it is the one limited space in his life where he can be openly gay. "I'd go crazy if it wasn't for It'aewŏn," he once put it, and the district, he thinks, is his safety valve that allows him to let off the steam that builds up during his busy weekdays. While this conceptualization is surprisingly similar to the way some GIs make sense of camptown spaces (as we have seen in the last chapter), Bong-hee does not necessarily think of soldiers as potential comrades to be encountered in his neighborhood. In fact, Bong-hee repeatedly told me that the one disturbing factor in "his" It'aewŏn was "straight" servicemen invading the space of "Homo Hill."

Occasionally, trouble would erupt on "Homo Hill" because an intoxicated soldier looking for "Hooker Hill" got lost on his way. Additionally, quite a few heterosexual GIs actually came to "Homo Hill" just for the music, and, to make matters worse in the eyes of Bong-hee, they often brought their girlfriends along to the gay bars and clubs where the DJs usually played techno rather than the R&B and hip hop that dominated the "straight" clubs of It'aewŏn. Bong-hee, who was in a relationship with a Korean American man and was dreaming about moving to the USA or to Europe in the not-too-distant future, was at times extremely irritated by the soldiers in "his" clubs: "They come here, they drink, they get loud and aggressive, they think they rule the place."

During his university years, he had been involved with a student group lobbying against US bases, and time and again they would demonstrate for the withdrawal of US troops in front of military installations. He first learned to fear American soldiers as a child, he explained, because his parents' house, located in a small town close to the Demilitarized Zone, stood on the very margin of a red-light district catering to US military personnel. Wanting to get away from all of the drinking and debauchery that he witnessed there as a kid, the irony does not seem to escape him that he now has to begrudgingly share his most beloved space with men very similar to those that he was so afraid of when he was a child.

It'aewŏn *suspense*, the at times rather uncomfortable symbolic positioning of the neighborhood between hyper-masculine violence and hedonistic pleasures, also entails another degree of *suspension* that

Bong-hee's story hints at—a temporal one in the sense of a "time-out" from ordinary life, which may also allow for some unexpected fraternal bonding. Perhaps It'aewŏn can be understood as a kind of liminal space that creates unlikely opportunities for those who enter it; consequently the emergence of a highly fractious spontaneous *communitas*[8] (Turner 1967, 1969) among the military and civilian male visitors of It'aewŏn may be interpreted as one result of this liminality. Joo-hwang, the former KATUSA, for instance certainly liked to reminisce about the many evenings he spent partying with GIs and a wild assortment of other actors that they randomly stumbled upon during their drinking tours of the neighborhood, even though, on other occasions, he would be quick to verbalize his misgivings about US soldiers in the area.

A Really Violent Bunch?

"I do have stories that don't involve violence. They are just not as good." This is what Eric said to me, a 25-year-old GI, a statement marking the end of a series of rowdy stories involving brawls breaking out in bars, random make-out sessions with Korean girls, disputes with strippers, prostitutes, and pimps over money and services, or fights between the soldiers themselves that usually resulted in bruises or broken bones. One story, however, that did not involve such a bloody punchline, was about how he had come to an inner-city bar for the first time, rather than just hanging out in a ville or in his favorite area, It'aewŏn. He and a buddy of his had walked into a hip bar in the ritzy Kangnam area that they had heard about, and the moment they stepped in, the entire place went quiet, with all eyes resting on them. Easily identifiable as US soldiers by their haircut and general physique, there was no place for them to hide that day. It turned out that a group of Korean peace activists, among them also one rather outspoken American one, were at the bar that day as well, and the hostility toward the two GIs walking into "their" bar was tangible in the air. They left again after one quick beer to avoid trouble. "You know," Eric summed it up for me in a frustrated tone:

> it's like that—we walk through the door and everyone knows what we are. Our haircut immediately gives it away ... And I felt like introducing myself to everyone: "Hey, I'm Eric from the States, and I haven't even finished high school, but they sent me to Korea anyways. So you wanna be friends?"[9]

I first meet Eric and his buddy Paulo on a rainy night in It'aewŏn, while they were in the company of Karen, an English acquaintance of mine. Karen herself had got to know Eric a while earlier at the end of a night of partying in this district, too. "It was my first weekend in Korea, and I was in It'aewŏn with some friends," Karen would later explain to me:

And that was my first experience of … this whole separation of people in Korea. 'Cause I stayed in this area where there were only Koreans. […] And then I ended up in It'aewŏn that night, me and a German friend, and it was crazy to see all the foreigners in one place. And she had brought some friends along, and they were all like … the girl was an English teacher, and the guy was a soldier. And I think that was the first time it really struck me what this place is all about.

On the subway a while later, yet another US soldier unknown to her, would sit down next to her, introduce himself as Eric, and explain to her that he had just arrived in Korea a week ago himself. She felt he came on to her too strongly that evening, though, and was a little concerned about actually giving him her phone number, so she decided to hand over her email address instead. After some email exchanges, she decided to meet Eric for coffee after all, only to discover that he was a decent guy. Eric would tease her in front of me about her reluctance to meet up: "You were incredibly rude to me, in fact!"

I was supposed to meet up with Karen, Eric, and their friend Paulo in front of a coffee shop on the main street of It'aewŏn. They would be coming there by car, I was told, as their base was located about an hour south of Seoul. Standing by the side of the road, I suddenly found myself approached by a tall, muscular Latin American, who swiftly walked up to me, grabbed me by the shoulder and said "Come with me," while he lightly pushed me toward a vehicle nearby. "Now, you really have to work on your abduction skills," Eric told Paulo the moment we got into the car, with Karen grinning back at me from the front seat, saying, "She didn't look too surprised." "Well, you gotta understand," Paulo said. "I could have used more force in this abduction, but I have only half a year left in Korea. I just don't want to create another one of those GI incidents that they can plaster all over the news here."

The club the two guys chose to take us to was a fancy looking Latino bar, located in the more up-scale restaurant section of It'aewŏn. Fake palm trees lined our way up the creatively decorated stairs and soon we found ourselves seated at a nice table by attentive bar staff. On the way there, we had run into several of their buddies, who were just about to enter H., a rather seedy bar popular with GIs. Paulo made sure to quickly usher us past his friends. "I figure you don't want us to hang out with *them*," I said, and Paulo shook his head, "Not tonight, no." "This one guy you saw, Jim," he explained, "he is a decent enough friend when he is sober, but once he's drunk he does have a bit of a God-complex." "What does that mean?" I wanted to know. "Well, he loves to stand in the middle of heavy-trafficked streets, that sort of thing, you know?" Nowadays, Paulo added, instead of going to the infamous GI bars and clubs of It'aewŏn, he himself preferred to go to more quiet places anyways.

Both Eric and Paulo joined the military at the age of 18, essentially growing into adult manhood inside the military. Paulo said:

> This is one thing you have to understand about the army. If you behave right, if you pull your own weight, if you try to fit in, it's a brotherhood, it really is. But every once in a while, there's a guy who wants to make trouble, mess with people, who slacks. Then there will be punishment. People will take care of it themselves, and the superiors will look the other way. That's how it's always been.

To prove their point, Paulo and Eric told us a series of stories now—one time, a guy had a stone shot at him with such precision that it dislocated his shoulder; another time, a soldier was thrown down the stairs and, finally, another offender that they knew had his head banged into the wall during shower time. During all these occasions, they told us, these acts of violence emerged out of group consent that this person needed some corrective discipline to come his way.

Physical injuries resulting from such incidents and from random fights breaking out, it seems, would become a daily part of life for them once they joined the army; yet from all that Eric and Paulo told us, a certain normalization of violence had already crept into their lives before that: "I got into my first knife fight as a 2nd grader," Eric, who hails from a military family in South Carolina, would say to me, "and in the army, of course, things just kind of continued that way." "Americans," he then added in a pensive tone:

> we are a really violent bunch. I only really understood that once I came to Korea, I think. We're good, peace- and fun-loving people most of the time, but then there always comes this moment, when someone snaps out of the blue. And when the fighting starts.

"With whom do you fight?" I asked, "With Koreans?" "No, more among ourselves. But when I hear people talk shit about America, yeah, then I do snap."

While Paulo grew up outside of the US, for Eric it was his very first time abroad, and he had some rather intense reactions toward South Korea and the at times very outspoken anti-Americanism that he experienced on a frequent basis.

Korea? When I first came there, I was so amazed by the country, it was so beautiful and interesting, and the people all seemed so friendly. After a while that really wore off, and I started to see that folks are actually quite cold to you. The old Koreans are ok, they still very much understand why Americans are in the country and they are often very grateful for us being here. But the young ones, that's often a different story. They have a rather skewed vision of why we are here, I guess. On 4 July, for instance, they celebrate, too, and sell us all the typical stuff, have no trouble taking all our money, and a few days later they go about protesting American beef. Well, I say, fuck you, why don't you first take care of all those cases of bird flu that you had last year rather than bitching about American beef? That thing I just don't get.

Eric was at that time dating a Korean woman whom he got to know in It'aewŏn, and he spent much of his free time with her or out drinking with his friends. He and Paulo also loved the alternative hangouts in Kangnam and the calmer corners of It'aewŏn; sometimes they went to Sin'chon, too (a student district adjacent to Hongdae)—but Eric was trying to stay clear of Hongdae itself because of the fact that this entire entertainment district had been declared off-limits to GIs after the 2007 rape case that involved an elderly Korean woman and an American soldier.

It happened one and a half years ago. But still they bring it up in the media here all the time. Yeah, I know, it was awful and shitty, but then … It was just one guy. And they make assumptions about all of us based on that … Hey, a Korean guy just shot dozens of people in the US, but do I hold that against Koreans in general?[10]

Paulo, on the other hand, loves Hongdae, and readily admitted to hanging out there whenever he could. He particularly liked some of the hardcore and punk venues in the area, where he had spent some great nights in the past. When he first came to Korea five years ago, he told us, he was first stationed further away from Seoul, near Ŭijŏngbu—the most horrible time for him, he says. He was new to the military, very skinny, tall, and awkward, and only had this tiny ville for his entertainment, where people would routinely get drunk and go after the Filipina entertainers. "First I got trashed every night like everyone else. Then I started going to the gym instead. That saved my mind." Things got better for him when he was finally relocated, stationed closer to Seoul. Every single minute of his free time he would now try to get out, get away from everything related to military matters, get on the subway and get off at random stops to see what this country was all about.

Like Eric, Paulo was bound to leave Korea over the next half year. Almost six years of his life he had spent in Korea, but he said he was now certainly ready to leave. "You know what it feels like when no one ever wants to sit next to you on the subway?" he asks me. "When people swear at you for no reason?" "It's true," Karen jumped in, "I wouldn't have believed it, but I've seen it too many times myself when I was out with him." No taxi would ever stop for them, Karen added, she was usually the one who got the cab for them, with the driver grumpily letting the soldier into his vehicle after he had stopped for the female foreigner holding up her hand in the street. "Yeah, it's time for us to go," Eric added with a bitter smirk.

It'aewŏn('s) Freedom

> Ta allyŏ chugaessŏ, ta malhae chugaessŏ
> / I'm gonna tell it all, I'm gonna tell it all.
> Saeroun saesang kŭgosŭl malhaepwa
> / Tell me where the new world is.
> (U.V., It'aewŏn Freedom)

A year before the K(orean)-Pop wave turned into a worldwide tsunami that hit clubs around the globe with a song called "Gangnam Style," another tune with lyrics that glorified a Seoul neighborhood had already become a huge success, albeit one that was restricted to South Korea: "It'aewŏn Freedom." In the famed music video, three Korean singers (that is, the comedy duo U.V., and Park J.Y.), dressed in black leather and equipped with fake Afros, can be seen dancing in a studio set of an entertainment district. The (more popular) entertainment districts of Kangnam and Hongdae are overcrowded and uninteresting, they sing, only It'aewŏn opens the door to new worlds for its visitors.

In addition to these three main protagonists a heavily made-up young Korean woman and a black man are also featured, who plays in turn African migrant, exotic tourist, and GI. Much of the humor of the video is derived from the fact that the woman is repeatedly rejecting the advances of the three local men, whereas the foreign character is increasingly seen as undermining their masculine assertions. While the Korean singers try to act tough, at the same time they are continuously frightened out of their wits by the sheer physical presence and size of the foreign male. The last few scenes of the music video then move directly into the streets of It'aewŏn, where the three are first seen cruising through the neighborhood in a limousine and then engaging in dance-offs with a crowd of partiers, many of whom are made up of intoxicated, cheerful foreigners.

It is an image of It'aewŏn with a long tradition that is being recycled here. Harking back to earlier histories of outright US domination, and the notions of freedom and liberty it paradoxically engendered, it reminds us how the district has always found itself critically suspended between different powers since the second half of the 20th century. With local leaders uneasy about the potential spread of "American things" that were seen as corroding the national spirit through potential entry gates such as It'aewŏn, the "ghettoization" of those viewed as having the potential to advance Western tastes was a direct outcome. I have argued that these old tensions created by such informal containment policies—which only supported the rapid fermentation of the very forces that were supposed to be kept under control—can be felt to this day. Just as the lyrics of "It'aewŏn freedom" were shouted across dance halls all over Seoul in 2010, to this day new forms of social, sexual, and perhaps also political engagement occasionally spill from It'aewŏn into other areas of Korea.

In this chapter about a peculiar Seoul entertainment district, I have sought to achieve two things. First, I have paid attention to the complicated interactions between local men and US soldiers within the terrain of It'aewŏn. The multiple performances of masculinity to be found here interact, compete with, and contradict each other in often unexpected ways in the mundane areas of the entertainment area. The experiences of two US soldiers, Eric and Paulo, and of two young Korean men, the former KATUSA Joo-hwang and the gay office worker Bong-hee, who all hold this neighborhood in high regard, have served as examples for some of the highly fractious masculinities that come into existence when US soldiers and South Korean men meet in this neighborhood. While the prior focus of feminist writers on the contentious sexual encounters of GIs and Korean women has shed much light on the perilous legacies that have been created by US bases in this country, this picture necessarily remains incomplete if we disregard the voices of US servicemen, and those of local men relating their unique experiences with the US military in South Korea.

Second, I have attempted to make sense of It'aewŏn as a place-making project that is ongoing, contested, and creating ever new boundaries that local and foreign actors need to navigate around, with the presence of the base having turned the area into an "impossible incubator" (Kim P. and Shin 2010: 203) for cultural and social (ex-)changes on the ground. Thinking of It'aewŏn's territory as a project that is defined by fierce competition over what the neighborhood might actually stand for, the only thing that holds It'aewŏn together as an imagined terrain at the end of the day may just be a vague notion of "a good time" that functions as the lowest common denominator for all of its actors. However, the suspense of It'aewŏn (an affective dimension created by the difficulties of governing such a diverse space through multiple forces that make competing sovereignty claims) engenders a terrain that certainly allows for some spatial and temporal suspension from people's ordinary life, and gives room not only to masculine competition, but also to some fleeting fraternal bonding between an unlikely cast of actors.

In conclusion, it has to be noted that the dreams, visions, and liberties that are often attached to the name of the neighborhood are nowadays less associated with ideas of an American Dream but are more firmly connected with the notion that foreign eroticism of all kinds can indeed be consumed in the here and now. With the temporary eradication of differences that follows out of the (at times playful, at times very real) concealment, clash, and mixing of identities taking place in It'aewŏn, the district becomes a rather unexpected entry and exit point for those seeking to engage with people they would never encounter in their ordinary lives. It'aewŏn's suspense, then, is also rooted in the fact that a visitor to the area may vaguely expect to party with US soldiers, but often finds himself hanging out with Colombian migrant workers, Korean transgender sex workers, or Nigerian traders instead. It'aewŏn today, more often than not, entails the promise of erotic consumption and fulfilment in an ever fleeting present within a specific terrain of Seoul, rather than a relegation of hopes and desires into the distant, yet still imperious, landscape of the USA. Within such an emerging regime that in its wake threatens to thoroughly commodify the district's infamous reputation (while at the same time increasingly purging the "women of ill repute" who have quintessentially shaped this notoriety), imaginaries of the dangers and joys that the US military brings in its wake are still curiously close associates.

6

Demilitarizing the Urban Entertainment Zone?

Hongdae and the US Armed Forces in the Seoul Capital Area

Spoiling the Show?

> There are so many foreigners [in Hongdae]—some of them are really bad people. American soldiers, you know. They misbehave. It's not safe for women there at night. I don't like to go there ... There are also decent foreigners, but the bad ones really spoil it for everyone. (Hyo-jin, 25-year-old student)

A group of ten friends and acquaintances had already assembled at Club O., a large Indie venue at the heart of Hongdae, when I realized that the evening could possibly turn into one tense occasion. Karen, an English teacher from the UK, had just sent me a text message that she was on her way and would bring a few soldiers along. I looked around, sizing up the assortment of Korean and foreign activists, leftists and punks gathered in this venue. A charity concert was just about to begin, with the goal of collecting money for overseas survivors of the "comfort women" system. My friends knew about my research and were supportive of it, but actually hanging out with GIs in their free time, and on such an occasion, would be another matter altogether for quite a few of them.

When Karen arrived at O., I was relieved to see that Karl was among the guys she had brought along—he was an Asian American soldier who was already well known among some of the people assembled, and who had, after a bit of reluctance, been welcomed into their circle. The two other guys, Tony and Steve, who were both white Americans in their mid 20s, seemed to be respectively uneasy and excited to be at Club O. "What an amazing place," exclaimed Tony, who, like Steve, was wearing a baseball cap to hide his short cropped hair. Karl, I knew, always smirked at such attempts by soldiers to disguise themselves on their nights out in Hongdae. He himself, at any rate, was never taken to be a serviceman by Koreans who did not already know his profession. And indeed, the guys seemed to be wearing their caps to no avail tonight, either—from the corner of my eye, I saw that three of the people who had sat with us earlier decided to move away from our group now.

Tony, a big and boisterous man, immediately started a loud conversation with Karl about his tour to Iraq that Karl had finished the year before. The softer-spoken Steve, who had just attempted to explain to a Korean friend of mine that they were all English teachers, looked rather mortified when he heard their talk. Repeatedly, he shushed Tony and Karl to get them to keep their voices down, even telling Tony in the end, "Dude, you are being disrespectful here" as the official event had already begun. Practically the moment the young soldiers had sat down with us, a Korean American woman had taken the microphone and started to read out a text on the comfort women issue that ended with a thunderous "End all Wars!" Tony looked at me for a moment now and then said with a smile: "Right. Time for a beer, what do you say?"

While It'aewŏn's emergence as an entertainment district is inextricably linked with the US base it is situated next to, the history of Hongdae in central Seoul is firmly connected to a learning institution in this area that has given the district its name—Hongik University (Hongik Taehakyo, also known by its abbreviation Hongdae). Founded in 1946 and well known for its highly regarded arts departments, this university would see a significant expansion of its programs and facilities from the 1980s onward,[1] which fundamentally changed the character of the originally quiet residential area around it that is today part of Seoul's Map'o district.

Attracting several private art institutes to the neighborhood in the early 1990s, eventually shops, stores, and a handful of bars and live clubs opened as well (Cho 2007; Chun 2002; Lee M. 2004: 70). The appearance of live venues, in particular, was quite remarkable: after the heyday of live music in It'aewŏn in the 1960s and 1970s, rock bands performing in front of an audience had gone out of style by the 1980s. It was only in the 1990s, and especially in Hongdae, that this entertainment concept regained some ground in South Korea. Of crucial importance for the renewed interest in live music performances in Hongdae was the 1994 opening of Club "Drug," a punk-rock location that would be among the first of many venues to promote the homegrown music of local alternative bands in this area (Cho 2007:47).[2]

The late 1990s and early 2000s then saw the emergence and rise of yet another type of establishment in Hongdae that many of those loyal to the live music scene viewed with much contempt: one dance club after another opened, playing hip hop, techno and other forms of dance music, and attracting ever new types of people to the district. Suddenly, "there was an increased clustering of people engaged not only in fine arts and music, but also in film, publication, design, advertisement, and internet development" (Lee M. 2004: 70), a combination of actors that necessarily left its imprint on the neighborhood, too. The opening of these new clubs would make Hongdae decidedly sexier in the eyes of many. For instance, in contrast to traditional "night clubs" that could be found in other parts of Seoul—usually expensive establishments in which businessmen could enjoy the company of young women who had to be paid for their services—these new types of clubs normally had a relatively low cover charge, had no female hostesses on the payroll, and generally left it to the visitors themselves to create a steamy atmosphere. In this way, the notion that short-lived sexual adventures free of charge could easily be attained in this neighborhood was born in the clubs of Hongdae.

The speedy gentrification of the district[3] that had occurred over the previous decade or two seemed nearly complete by 2007, when I first came to Hongdae myself. At the same time, the "old Hongdae" was an image that lingered on, and talk about how the "real" Hongdae had been "ruined for good" was certainly one of the favorite topics of conversation among many of the people I would meet there. I soon learned, though, that the push of new capital, lifestyles, and people into the area had not led to a complete displacement of its more alternative audiences. Young artists, musicians, and adherents of alternative lifestyles have de facto refused to let themselves be driven out of this space. Nowadays, they either cling to the Indie cafes, bars, and clubs left within the otherwise thoroughly commercialized areas, or simply loiter in the streets of the neighborhood if the weather permits them to do so.[4]

Around the same time that Hongdae's reputation as Seoul's party headquarters grew, a new specimen emerged on this urban stage that would, in the eyes of those who still imagined Hongdae as a haven for young left-wingers, cause even more trouble than the indifferent hordes of Korean clubbers: foreign residents living in the larger Seoul metropolitan area also started to come to the neighborhood at weekends, with their arrival quickly classified as a major source of annoyance by many. Yuna, for instance, a 30-year-old peace activist I occasionally met up with, had spent most of her 20s hanging out in Hongdae, and was certainly not alone in her views that foreigners were the culprits who brought about the demise of Hongdae. Ten years ago, she told me, the neighborhood was not only the center for Seoul's alternative music scene, but also a place for dissident thought and action. But "then all those … sorry … all those foreigners started to come. It totally ruined the independent culture, instead it just became about clubbing and corporate shops and restaurants."

Among the—predominantly Western—foreigners who frequent the neighborhood nowadays, one group, in particular, would soon be singled out as fundamentally responsible for the dissolution of the alternative district. US soldiers, seeking to leave behind the claustrophobic camptown areas designated for them, had started to come to the area *en masse*. They were drawn in by the free-spirited atmosphere, the large concentration of entertainment facilities, and, in all likelihood, also the availability of a large number of young Korean women who were curious about getting to know foreigners in the clubs and bars of the area. And with Hongdae's left-wing youth having been widely exposed to depictions of US soldiers as potential criminals throughout the years, US servicemen were now judged to be *the* prime offenders bringing their favorite neighborhood down.

In what is to come, I shall outline yet another materialization of the at times highly conflict-ridden encounter between US soldiers and civilians in the vast urban space of Seoul. What kind of (predictable or surprising) repercussions did the appearance of significant numbers of US servicemen have in this entertainment district that was understood by many to be *the* territory of dissenting young people? Some remarkable similarities and differences can be made out between It'aewŏn and Hongdae in the ways that the presence of US servicemen in these neighborhoods has been dealt with. As laid out in the last chapter, some of the elements of violent imaginaries (which emerged out of a political project in the 1980s and 1990s) have been incorporated into the very way the neighborhood of It'aewŏn is being territorialized. It'aewŏn's stakeholders have to some degree managed to capitalize on the bad reputation of camptowns, insofar as they have turned the notion of It'aewŏn as a sinful terrain into a tool that allowed them to attract new Korean audiences, too. In Hongdae, however, quite a few of the district's visitors, together with media commentators who have tried to make sense of this neighborhood, have chosen familiar negative depictions of US servicemen in order to actively keep American soldiers out of Hongdae, while simultaneously making moral claims about an area that had seemingly lost its old essence by becoming too entangled with foreigners.

If a territory, in line with Brighenti's (2010) understanding that we introduced in the last chapter, is indeed a set of practices that allows participants to connect the past with the present, read the signs of the landscape they find themselves in, and share a particular space in a purposeful manner, then Hongdae, just like It'aewŏn, is a heavily contested corner of Seoul that requires much everyday maneuvering by those engaged in it. And while overt forms of politics seemed rather inconsequential to most of the people I met in It'aewŏn, Hongdae is at heart a "leftist space" to many of its visitors, which has made the arrival of GIs on this urban stage an exceptionally explosive issue. In Hongdae, too, violent imaginaries have been turned into a social practice among local actors that is essentially about (re-)claiming a territory. But in this part of Seoul, these place-making struggles are also turned into a contestation over who gets to call the shots in the major processes of urbanization (see, for instance: Harvey 2008, 2012) that are constantly reconfiguring the densely populated space of Seoul.

Before attending to how various people occasionally come to a clash over the presence of GIs in Hongdae, in the next section, I shall first give a more concrete sense of this district's diversity by taking the reader on a quick tour through the neighborhood, and by presenting a few impressions of how the debates around US servicemen in the 2000s have played themselves out locally. In the following two sections, I will then look into two rather interesting occurrences that only tangentially involved US soldiers, but that were still *made* to be about their presence in the neighborhood. The first one—a scandal over a party organized by foreign English teachers in late 2004—directly touched upon an inconvenient matter that has preoccupied several generations of Koreans since the arrival of US troops on the peninsula: the issue of local women voluntarily engaging in sexual relationships with foreign men. The second occurrence involves a case of public indecent exposure by two young Korean punk musicians, an event that was taken as an(other) opportunity by the Korean media to paint Hongdae, where these young men primarily spent their time, as a neighborhood full of foreign-bred social evils: "Hongdae," one news reporter exclaimed at that time, "is now an area hot with youthful passion that has degenerated from being mixed up with foreigners" (quoted in Koehler 2005c).

In the final two parts of this chapter, I shall focus on the Korean milieu that these two punk offenders emerged from. A group of young punks I encountered in Hongdae was exceptional in the sense that they were not only outspoken about US soldiers in Hongdae but also held anti-militarist convictions that went beyond the conceptual horizon of most other anti-US bases activists I encountered. They tended not to shy away from addressing previously unassailable areas of South Korea's homegrown militarism, such as mandatory male conscription. Paradoxically, this group of people was often forced begrudgingly to share their favorite hang-out spaces with GIs who were visiting the same outdoor areas, music venues, and bars that they also liked to gather in. In this context, the participation of some of the punks in the extended struggle surrounding Taechuri—a small village south of Seoul that was partly destroyed in 2006 to make space for the expansion of a US base nearby—led to discussions over the US–Korea military alliance between "Hongdae GIs" and these young anarcho-punks.

Hongdae's Forbidden Fruits

The lockers at Hongdae subway station are usually all occupied, the little containers filled with colorful stockings, short skirts and high heels, whose owners come to collect them only on the weekends. Young women—first dressed in plain jeans and T-shirts—emerge from the subway gates in the early hours of the evening, grab these clothes and then head for the public restroom nearby to change into their Hongdae outfits. A space close to the mirrors is hard to come by on those occasions—too many women shuffling around, checking the state of their hair, clothes, and shoes, while putting the finishing touches to their make-up. Fully transformed in the end, they make their way up the stairs and head into the night, only to return to the restroom many hours later, in time to get changed into their "decent" clothes again before the last subway arrives that will take them back home to their parents.

Following the large numbers of young people toward Hongdae's subway exit number six can turn into quite an ordeal on such weekends—at times, it may take a visitor a good 15 minutes to cover the few hundred meters that lead you outside the station. The experience of overcrowding usually does not end with one's emergence into the open air; instead, the next challenge is to make your way past the hundreds of young people who are waiting in front of the exit for their friends to arrive. Occasionally, groups of Christian performers use this area to sing about Jesus to those who are gathered there, and young men push their way through the crowd, handing out leaflets advertising yet another newly opened club.

Walking into a side street filled with little stores selling make-up, cellphones, and accessories, then past a long row of busy barbecue restaurants where large numbers of students enjoy their grilled meat with nearly endless supplies of *soju*, you will soon arrive at a major avenue that leads up to Hongik University. Near the rather imposing main university building, the houses are all filled with up-scale shops, restaurants, clubs and bars that have flashing neon signs attached to their entrances in an attempt to lure more people in. If you keep to the right now, you will find another side street lined with little boutiques that sell "Hongdae style" fashion, next to quaint little coffee shops where any beverage will cost you 6,000 won ($5) or more. Some of the most popular hip hop and dance clubs are located nearby, and at one prominent Karaoke bar that rises several stories high into the sky, you may catch a glimpse of people performing their favorite songs behind a glass-covered front: young Koreans singing in little see-through cubicles, dancing to a laser show of their choice. While still gazing at this spectacle, the visitor is being pushed forward by endless numbers of partiers—hipster boys wearing fedoras and checkered suits, punks with studded black leather jackets, gangs of Emo kids drifting by, a girl dressed up in a sexy cop costume, handcuffed to her most recent boyfriend for the length of the night.

Follow another side alley, and you will end up at the indisputable center of gravity of Hongdae: a little park that is known as the *Hongdae Norit'ŏ* (Hongdae playground). On warmer weekends, the so-called free market takes place here, with handmade jewelry, paintings, and clothes being sold during the day. Once the sun goes down and the vendors slowly disappear, the playground by no means grows quieter: street musicians gather large crowds around them, playing free gigs until the police arrive, B-boys are cheered on by those watching them display their skills, groups of hippies assemble for a drumming session, and sometimes makeshift theater performances are put on during warm summer nights for the amusement of people walking by. Kids of all subcultural denominations loiter at the curbs of the sidewalks, make themselves comfortable on the swings or slides meant for children, or simply sit on the dirty ground, where they pass around drinks and food that can be bought cheaply at a convenience store across the street. A couple of times a night, the famous "makkŏlli man" makes his rounds, a middle-aged street vendor who drags his cart filled with home-made rice wine (*makkŏlli*) through the streets, and entertains people with free samples of the beverage and a small dose of his outgoing personality—both to be enjoyed in this outdoor location, the cheapest and most diverse Hongdae venue of them all.

Mike is a bear-sized, heavily tattooed US soldier I occasionally run in to while hanging out at the Hongdae playground. He is with another, equally intimidating looking military guy tonight. They just returned from an Italian restaurant nearby, Mike says, and loudly complains to me now that the waiters there had continuously given them dirty looks throughout their meal. So eventually he went over to say hello and talk to them for a while, but the conversation did not go very well, he says: "You know, usually people just start to revise their opinions a bit when they see, oh, this person is actually not that bad ... Not those guys, though, oh no." While chatty Mike readily tells anyone who asks that he is a soldier, his friend announces to me tonight that he is in education. I size him up for a second, and then I let it go; not the type of guy to call a liar to his face. He tells me he is trying hard to find a Korean girlfriend these days, as he would never be able to learn this language otherwise; then we talk about Europe. He tells me he has European ancestry and asks me whether men there look similar to him, and if I believe that he would blend in with the crowd. He certainly does not blend in here in Seoul; at the Hongdae Norit'ŏ, people sitting nearby keep glancing at him and his friend in a curious fashion.

From It'aewŏn to Hongdae, it's a 30-minute ride on Seoul's subway. From Tongduch'ŏn in the north, or from the P'yŏngt'aek area in the south, however, it will take you a good one and a half hours via public transportation to get to this inner-city neighborhood—a substantial commute that brings together a number of places that, in the eyes of many, could not be further apart. Bosan station, nearly the last stop on Seoul's subway line no. 1, for instance, connects Tongduch'ŏn's heavily contested camptown with the glitzier city center of Seoul; a remote subway station that was only opened in 2006, and has allowed soldiers to access the capital with greater ease.

The swift and ongoing integration of such marginalized areas outside of Seoul into the capital's public transportation networks is but one symptom of the breathtaking urban development that has been remaking the wider Seoul region, where, because of a severe lack of affordable living space within the capital itself ever new territories have been incorporated into one gigantic, sprawling urban space. This process of urbanization that is affecting much of the northern half of South Korea has also inadvertently undermined an informal containment policy targeted at US camptowns and their inhabitants that we have discussed in previous chapters. The quasi-"ghettoization" of foreign soldiers proved to be a strategy that was entirely undermined by Seoul's dramatic ascent as a megacity during the latest era of globalized capitalism.

The increasing number of GIs who started to come to Hongdae in the early 2000s, however, caused much concern among local club owners in particular, who were afraid of losing their Korean clientele because of the arrival of these new guests at their venues. The soldiers, one rumor at that time went, had started to charter entire buses to get from remoter base areas such as Tongduch'ŏn to Hongdae, and "GIs were becoming such a common sight in Hongdae that Koreans started calling it 'Hong-itaewon'" (Chun 2002). With anxiety over US servicemen in Hongdae massively on the rise, in late October 2002, US soldiers coming to the neighborhood were faced with signs at various clubs, notifying them that they would be barred from entry to these establishments:

> The first thing you now see at the threshold of the 10 [Hongdae] clubs is a yellow, 60-by-45 centimeter sign in English. In bold, black capitals it says: "We sincerely apologize, but due to many previous bad experiences, GIs are no longer permitted to enter Hongdae clubs." The letters in "GIs" are bright red. Right next to the sign is usually a red sign that warns, again in English: "Things not to do at Hong-dae clubs." The list provides hints as to what the "previous bad experiences" may have been: drugs, fights and sexual harassment. (Chun 2002)

This ban on US soldiers, incidentally, happened just a few months after the accidental death of Shim Mi-sŏn and Shin Hyo-sun, two teenagers who were killed by a US military vehicle in the summer of that year. In the fall of 2002, expectations were running high in Seoul and beyond that the drivers involved in the death of the two schoolgirls would receive a jail sentence. When the two were then cleared of negligent homicide in a US military court in November, protests in South Korea quickly spread, bringing tens of thousands of protesters to the center of Seoul.

In the midst of such public turmoil over the US military presence in South Korea, the atmosphere that US soldiers stepped into whenever they entered Hongdae in late 2002 was indeed very tense, and on occasions bordered on the hysterical. For instance, a wild, unconfirmed rumor was making the rounds at the clubs and bars at that time, claiming that a US serviceman had stabbed a young Korean woman at a club that was popular with GIs: "According to the rumor, the woman died immediately, and the GI was handed over to the US Army and nobody knows if he was punished" (Chun 2002). Hongdae club owners, together with concerned student activists from Hongik University, then held meetings to discuss the increasingly controversial presence of US military personnel in

Hongdae and jointly decided on the ban: "The students say they have seen enough of the 'arrogance of GIs'. And that the soldiers 'have an attitude that Koreans should be grateful that the U.S. Army is there to protect the country'" (Chun 2002).

The US Armed Forces, alerted by the furore in Hongdae over their soldiers, reacted within a month and, due to "force-protection concerns," put the entire neighborhood off limits starting from 2 December 2002. This action meant that, for their own safety, all troops and their dependents were to stay out of the area from 9pm to 5am. The order would be lifted only on 1 May 2006, after "Korean National Police, U.S. military police and force-protection officials [had] conducted a combined threat assessment, officials said" (Flack 2007). However, seven months later, when Army Private Geronimo Ramirez raped a 67-year old Korean woman at the end of a night of partying at Hongdae, in the midst of the ensuing public storm of outrage the authorities were forced to revise their position again.

The ban on US soldiers going to Hongdae, which has once again been lifted in the meantime, was still in place during the period of my field research in 2007–9, yet US soldiers frequenting the area were not exactly hard to spot. Having grown somewhat cautious in the way they acted while in Hongdae, servicemen tended to stay away from certain clubs and venues where they knew they could run into trouble, and often used baseball caps to cover up their short-cropped hair. Just like Mike's friend and Steve, whom I met at Club O., many would tell you at first that they were English teachers if asked about their profession. Paradoxically, the fact that US soldiers were not exactly invited to join the ongoing parties made this space even more attractive to some of them. Karl, who was a 26-year-old Asian-American serviceman, explained his passion for this neighborhood to me in the following way: "It's very, very far away from It'aewŏn, and all these other areas for the military. It's more like a college town atmosphere. 'Cause there actually is that university right next to the place … It feels like home."

Karl, who holds a humanities degree from a US college back home, was often thinking back warmly to his formative student years before he had joined the army. Like other GIs I met, Karl was also keenly aware of the many debates surrounding the US Armed Forces presence in South Korea. The feeling of unease he felt when out in town was often heightened by the fact that Koreans often mistook him for a local man, and at times started to berate him when he spoke in English to them. He had only recently arrived in Korea after a deployment in Iraq, and would occasionally talk about the various signs of post-traumatic stress disorder that he observed in himself since his Middle Eastern tour came to a close. While hanging out in Hongdae, he was usually a cheerful and amicable companion, but every once in a while I saw him get terribly drunk, which would inevitably lead to hostile situations with inebriated Korean men. A few months after I had left the country, I heard from friends that he had gotten into a bad fight with a group of young Koreans at the Hongdae Norit'ŏ because he had accidentally antagonized them in the early morning hours, and that he stopped coming to Hongdae afterwards.

Figure 6.1   The Hongdae Norit'ŏ (playground)

Yanggongju Revisited: "Are Western Bastards That Good?"

I met Min-ho, a 32-year-old Korean office worker, in a popular foreign hangout in Hongdae, where we struck up a conversation at the bar. I had just lost sight of Suzie, my 23-year-old American acquaintance, who had taken me to this venue. "Popular with GIs," she had explained beforehand, but once we arrived at the club, I saw that the place hosted a very mixed crowd of young Koreans, civilian foreigners, and a few men here and there who looked like they could be servicemen. English teachers like Suzie, who had graduated from a college in the Midwest just a year earlier and who usually spent her weekends in Hongdae, were very welcome in clubs like these. Just as in the bars of It'aewŏn, in this venue, too, English could be more frequently heard than Korean among the effervescent crowd.

"Are you Russian?", a Korean man addressed me in English. He sported a stylish leather jacket, washed out jeans, and one of those black-and-white fedoras that could be spotted by the dozen in the streets of Hongdae those days. "Why, do I look Russian?" I had grown a bit wary of being asked this question by Korean men; in It'aewŏn, it had usually proved to be a thinly veiled inquiry as to whether or not I was a sex worker. My new acquaintance just laughed, "Russian women are the most beautiful on earth, don't you think?" His name was Min-ho, he said; he spoke excellent English, and I asked him whether he had ever lived abroad. Never, he replied, but he adored foreign women, and listed the nationalities of a few of his ex-girlfriends now. Min-ho had learned how to speak English in the bars of It'aewŏn, he explained. He had even played in a band with a few soldiers for a while, that's where he learned practically everything he knows about foreigners, he added with a smirk.

After years spent partying in It'aewŏn, nowadays he preferred Hongdae, because "this is where the real parties are happening. I mean, just look at this place," he said, vaguely pointing at the foreigners and Koreans around us who could barely move forward, as the club was increasingly packed with people now. At this moment during our chat, a Korean woman, seemingly in her early 20s, who had been flirting heavily with a white, muscular guy with shaven hair across the bar from us, laughed so loudly at a comment of her acquaintance that the flow of our conversation was interrupted. We both glanced at this other couple now. "Look at that," Min-ho said to me without a hint of irony, "so shameful. All these Korean girls, they just come here with one thing in mind, to have sex with a foreign guy."

Min-ho's comment somehow cut to the core of many similar conversations I had had during field research. Liaisons between Korean women and Western men that began in Hongdae, I was to learn, were frowned upon by the most diverse set of people I encountered in Korea: young Korean men who often used strong words such as "disgrace," or "shame," when talking about this issue; a Korean housewife who warned her daughter in front of me to stay clear of Hongdae and all its scary foreigners; an It'aewŏn-based sex worker who complained that Hongdae's women were giving it up for free, thereby ruining the market for sex elsewhere; and Korean activist-types who told me that certain corners of Hongdae were off-limits to them because they did not want to have to share a venue with US soldiers and the "irresponsible" women who hung out with them.

The troublesome issue of local women freely engaging in sexual relationships with foreign men was also at the center of a public dispute that erupted in late 2004. The affair was triggered by an "advertisement" posted in the *English Spectrum*, a (now defunct) online forum that was popular with foreign English teachers working in South Korea in the early 2000s:

> Party humpers, Just so there's no confusion … English Spectrum and I will be hosting two parties at MaryJane's in Hongdae on BOTH the 14th (Friday night) and the 15th (Saturday night).
>
> Each party will be slightly different. On the 14th, it will be much the same as the last two; meaning some sex in the female bathroom, some late night dance floor grinding and partial nudity, mixed in with the addition of some clothes-allergic professionals who should be making a guest appearance that night.
>
> On the 15th, we will be holding an MC'ed Sexy Game Night. We will be selling drinks at an exceptionally low price from 9–12 to get everyone hammered prior to the games. From 12 to about 4 am, we'll play a bunch of team-oriented, guy-girl-guy-girl games, each for small prizes. This will be mixed in with a fair amount of dancing (hip hop and otherwise) intermissions.
>
> Both nights will be fun, but a little bit different. If you can make one or the other or both, please come and join the fun. The 15th will be a good time to meet new people and develop some interesting relationships.
>
> The Playboy

As a follow-up to this post, a series of pictures were uploaded to the same website, showing that the party had indeed provided everything "the Playboy" had promised. The pictures of partially exposed Korean women and half-naked white men partying excessively, doing the "bubi bubi dance"5 or making out with each other would then be leaked to the Korean news and blogosphere, after which the entire English Spectrum website was searched by Korean "netizens" for content that could be considered sexually degrading to Korean women.

And with comments such as "There is nothing good about Korea except that it's easy to sleep with the women and make money," or another user calling Korea "the Kimchiland where it's easy to score with the women and make money," they were certainly finding enough to keep the scandal afloat for the next few weeks (Koehler 2005a, 2005b). Such statements were used by the Korean media, bloggers and other internet users to depict male foreigners in the country as potential sex delinquents, whose favorite adult playground, Hongdae, had seemingly been turned into a pit of immorality, where the corruption of young Korean minds and the seduction of their bodies was the usual game of the night. Together with US soldiers, English teachers were now to become the new central figures representing the sexually corrupted and corrupting Westerner, who should be kept away from local women at all costs.[6]

On a very superficial level, there are some quite evident parallels to be made out between this particular scandal and earlier crime stories involving GIs, given that these English Spectrum parties, in the eyes of Korean netizens, also involved a triangle of aggressive foreign actors, local female victims, and an entertainment space of ill repute—three components that proved so central to the older discussions surrounding camptowns. And indeed, these resemblances were further highlighted in a rather crude fashion in the online discussions surrounding the young Korean females who were depicted so compromisingly in the leaked photos. With the women's names and addresses leaked, they suddenly found themselves exposed all over the Korean web and would go through much harassment and public shaming over the next few months, with some of their critics bringing these camptown resemblances into play. "Some online articles [...] said we were prostitutes, western princesses ["yanggongju"], and brothel keepers" (Choi 2005, quoted in Gusts 2009), one woman involved in the scandal said. Comments directed at the women labeled them as "Foreigner's whore! Why don't you shut down your club?" Another person queried: "Why don't whores like you just die quietly," while others asked them: "Whores, are Western bastards that good?" (Shin 2005).[7]

In this way, the notion of the "Hongdae yanggongju" was born—a term that establishes a curious link between a derogatory notion that was primarily used for sex workers employed in the GI clubs in the kijich'on areas of the country, and contemporary female visitors in Seoul's Hongdae entertainment district who occasionally chose to hook up with foreigners. As we may remember, the original usage of the word is indeed most tightly linked to camptown territories, and the very word "yanggongju," sociologist Kim Hyun Sook has argued:

> relegates Korean women working in militarized prostitution with foreign men to the lowest status within the hierarchy of prostitution. Since the end of the Korean War, this category has been extended to include Korean women who marry American servicemen (pejoratively called "GI Brides"). In postwar Korea, the epithet "yanggongju" has become synonymous with "GI Brides", so that Korean women in interracial marriages are also viewed as "yanggongju". (1998: 178)

In the tumultuous 1980s and 1990s, as we have seen, the "yanggongju" also became a central figure for the Korean nationalist left, whose actors established this social type as a potent (yet mute) symbol of a nation, understood to be constantly in danger of usurpation by powerful outside forces (see also Cho G. 2008: 89ff). This explosive baggage that came with the label of "yanggongju," then, was in a sense transported through time and space, and re-activated in the midst of the public commotion triggered by the raunchy party in Hongdae. Given this context, one can only concur with Wagner and VanVolkenburg, who have argued in a piece on English teachers in Korea that "the branding of these women as 'yanggongju' was highly significant in sparking a nationalist movement, as this label reproduced a well-known 'folk devil' of Korean society capable of provoking strong feelings of righteousness" (2012: 211).

The anonymous online attacks on the young women, which usually involved attempts to label the females involved as treacherous whores who have betrayed the nation, together with the more restrained and toned-down discussions by news reporters (who focused mainly on the aspect of "female sexual degradation" at the hands of foreigners) certainly paint a rather disquieting picture. The treacherous whore vs. misguided victim dichotomy that served as the predominant lens through which camptown women had traditionally been made sense of, was now taken out of its original context and moved to an inner-city entertainment district where the vast majority of actors engaged in sexual encounters without monetary incentives playing a role. In all its unsavory details, this scandal also shows how enduring the legacy of over half a century of camptown prostitution is in the public imagination, and how quickly old slurs and insults that emerged in camptown areas can resurface in times of perceived crisis, when Korean women's actions, in particular, are seen to be in need of rebuke.

"Sexual Harassment of National Proportions"

With foreign influences in South Korea for many decades mostly embodied in the shape of the GI, foreigners living in South Korea today hail from a number of countries, backgrounds, ethnicities, religions, and occupations.[8] While we may be "witnessing the very incipient stage of a process of gradual de-ethnicization of Koreanness, as Korean identity is being broadened to include plural cultures and multiple ethnicities" (Lee J. 2010: 19), the fear of losing this essence of "Koreanness" is also on the rise in a country where many citizens have long taken active pride in their extreme ethnic homogeneity. Within a "context of intensifying globaliza-tion of the South Korean economy and the consequent multiethnicization of its population" (2010: 19), spaces of consumption such as Hongdae, which increasingly function as *transnational* realms of hedonism and desire, have become signifiers to many for the broader capitalist processes that conquer ever new urban frontiers, irrespective of local histories.

These underlying concerns over rapid social change are equally visible in the debates around another scandal that broke loose just six months after the conflict over the "English Spectrum" parties. The incident, this time involving young Korean men, would once and for all prove to those already suspicious of Hongdae how far the putative corrosion of innocent youth had progressed in the streets of this neighborhood. On 30 July 2005, MBC television network aired its hugely successful live show *Music Camp* in its usual afternoon slot. This event, taking place once a week, sought to introduce new and promising live bands to its predominantly teenage audience. This time, a Hongdae punk band called RUX had been invited to perform a song in a studio setting in front of a largely female teen crowd, with the show simultaneously being broadcast live across the country.

RUX, a band that had acquired significant local fame in the gritty bars and clubs of Hongdae, had generously invited many of its street punk friends to come on stage together with them, I was told by Jil-Sung, a punk friend of mine I knew from the Hongdae Playground. Among others, two members of the punk band Couch had gladly accepted the invitation by RUX. "They got very excited immediately and started to think of ways to show those big TV guys what they really think of them and the shit music they promote," Jil-Sung explained. "They said they would drop their pants on stage, and we all laughed at the idea. Frankly speaking, no one thought they would really do it."

But indeed, when the moment came, the two musicians pulled their pants down and "exposed their genitalia while continuing to dance. The scene was broadcast for about four seconds. The two musicians [...] were arrested almost immediately after the show" (Kim T. 2005). Together with the two offenders, the lead singer of RUX, who had not actually exposed himself, "was also arrested for having invited the members of Couch, aged 20 and 27, on the show" (Kim T. 2005). The two Couch members would face a courtroom a few months later for their misdeed, and after prosecutors had initially demanded hefty sentences for the two of them (up to two years of prison time had been requested), they eventually walked away with 10-month suspended sentences (*JoonAng Daily* 2005).

The media, in the storm that broke out immediately after the incident, unsurprisingly had much to say about these "Punk rockers' privates in affront to Korea's 'bourgeois'," as one creative newspaper headline read at the time (*Chosun Ilbo* 2005b). For instance, the "unprecedented affront to Korea's conservative mores" (*Chosun Ilbo* 2005b) was decried as a form of sexual violation, as a *Korea Times* editorial on August 1, 2005 stated: "The independent rock band's behavior is inexcusable by any standard, as it was nothing but a kind of sexual harassment of national proportions" (quoted in Gusts 2005). Conservative newspaper *Joong-Ang Ilbo*, in another editorial, went for a similar line of argument by stating that "it's as if they've committed sexual violence against all viewers" (quoted in Gusts 2005).

Inadvertently adding fuel to the fire, an off-hand remark by the lead singer of RUX brought further unwanted attention to Hongdae as a place that bred deviant behavior. In an interview, he "said that the kind of performance seen during the MBC broadcast is common at clubs in Hongdae region, a hot clubbing district. 'We are free to perform there. Sometimes we break a guitar or bottles of beer', Won said" (Jin 2005). As a consequence, a policeman interviewed by the *Chosun Ilbo* promised that their investigation was going to be expanded "into unhealthy and corrupt performance venues and related businesses near Hongik University" (*Chosun Ilbo* 2005a).

A few days earlier, Seoul's mayor at that time, Lee Myung-bak (who would later be voted into the presidential office), had already chipped in and vowed to take action against the entire Hongdae Indie scene. His proposal was that blacklists of "indecent" bands should be drawn up, with those finding themselves on the list being permanently barred from performing at events organized by Seoul City or institutions related to it (Jin 2005). In a prompt reply to Lee's suggestions, politician Kim Hyun-mee from the leftist Uri Party accused Lee of trying to bring back the "discipline of Yusin," an open reference to the dictatorial times under Park Chung-hee, who famously had a vendetta against the alternative music scene in the 1970s (see Kim P. and Shin 2010). "I'm not sure if deciding who can and cannot be invited to performances under Seoul City is up to the mayor, but it's really an anachronistic and absurd order," Kim said. "[...] To call for a blacklist, label 'indie' culture subversive and try to restrict it is something the ghosts of the Yushin era would do" (*Chosun Ilbo* 2005c).

Finally, ten days after the indecent exposure, a *Herald Business News* article took the opportunity to remind its readers once more of the other great social evil besides raunchy punk performances that could be found in Hongdae: the sexual fraternizing between local women and foreign men. The clubs near Hongik University, the article claimed, were quickly changing into a foreigners' "paradise for hunting women." Recalling the (at that time recently revoked) ban on GIs in the clubs, the article's author stated that the good old days, when foreigners were not welcome in Hongdae, were now regrettably long gone. The article comes to the conclusion that:

> Hongdae is now an area hot with youthful passion that has degenerated from being mixed up with foreigners. As the recent act of indecent exposure by a punk band on live TV showed, the diversity and individuality of the area in front of Hongik University is nowhere [else] to be found. As the number of foreigners with more of an interest in [...] one night stands than in the music increases, there are many women coming to the clubs in search of "blue-eyed men." (quoted in Koehler 2005c)[9]

Figure 6.2   Live music show in Hongdae

Anti-Militarist Punks in Hongdae

On a typical Hongdae weekend night, it is quite easy to spot them at the Hongdae playground: a group of young Korean punks, many of whom are heavily tattooed and pierced, virtually all dressed in black. The girls wear torn stockings and black leather skirts with their heavy combat boots, the boys are typically clad in black jeans and hoodies that are covered in self-made patches they have sewed onto their clothes. Some carry bullet belts around their waists; others show off their studded leather jackets in this open-air space. And most certainly, whenever they have gathered at the Noritŏ, they can be seen drinking rice wine or the cheapest beer from a shared bottle.

While the young women of the group are typically still in high school or work in temporary jobs at the bars or stores nearby, the men keep themselves afloat with random jobs: working at convenience stores, bar-tending, or delivering food for various restaurants in the neighborhood. The hourly pay is usually small, but they typically get paid in cash at the end of a working day, which allows them to take their money straight to the Hongdae playground or a cheap pub nearby to hang out with their friends. Whoever is making a bit of money at any given time, so goes the ethic among them, is responsible for buying the alcohol that is to be shared with everyone else who decides to show up.

These young people's thorough disengagement from "Korea proper," where ambition and hard work are counted among the highest social values these days, gives them much time to maintain their close relationships with each other. Quite a few of them have dropped out of high school or are on the verge of doing so; only a few of them have even attempted to get a university education and none have succeeded. Their circumvention of a formal education may not strictly speaking be a choice, however: similar to the early punks in Europe and the US during the 1970s, most of the young men and women involved in the street punk scene of Hongdae come from economically disadvantaged backgrounds, which makes enrollment in exceedingly expensive Korean universities[10] difficult, to say the least.

Regardless of their lack of formal education, this particular group of punks has made a strong effort over the years to politicize themselves and their circle of friends. Appalled by the start of the Iraq War in 2003, some of them joined anti-war protests and in this way became acquainted with various left-wing activists. By befriending foreign anarchists (who were traveling through the country or working in Korea as English teachers), or through reading radical blogs and forums, they would inform themselves on the political matters that were important to them. They subsequently showed up at Esperanto classes, went to lectures and teach-ins organized by radical alternative learning groups, staged little protests against newly opened corporate coffee shops in Hongdae, and eventually got themselves involved in the struggle over the small village of Taechuri that was to make room for the expansion of US military base Camp Humphrey.

Jae-sŏk, who is in his mid-20s and among the oldest punks at the Noritŏ, functions as a role model for the younger people in this group. I had heard rumors before that he had actually grown up at the Demilitarized Zone (DMZ), so one day, I asked him about it, and he explained to me that he does indeed hail from a village so close to the DMZ that his family were under a military-imposed curfew every night while he grew up. The original village had been destroyed during the Korean War, and the government had eventually offered the land for free to repopulate this area. "Half of the [new] villagers came from the North," Jae-sŏk explained.

> They hated the North Korean government so they escaped from there, and many are still living in the village, being unable to return to their hometowns. My uncle is one of them. My uncle came down to the South to meet up with my father and aunt.

The other half of the villagers, he contended, consisted of poor people from South Korea:

> Back then [when the village was resettled], everyone was scared of the war breaking out again. So the only ones willing to move there ... half of them couldn't go back to their hometowns [in the North], and the other half just came for the free land. All of them poor people.

Following this, he explains to me that he got interested in a left-wing kind of politics due to his poor family background, too:

> When I see capitalism I find that it is really despicable. My family is poor; we have been going through a tough life. Of course I have been a lazy fuck, but [laughs] ... I see this poverty continuing through the generations. My mother is a really hard-working person. But as she continues to live a hard life, and as she works 365 days out of the year, it doesn't seem like we are getting any closer to getting out of this poverty. [...] When I got to know Jae-bong [a punk friend of his], I took part in these kinds of [political] conversations for the first time in my life. [...] And one day we watched TV, and the news reported that thousands of workers were gathering in central Seoul, that they would fight [the police] with iron pipes and bamboo spears. It was freaking shocking. Yeah, I could sense that they were mistreated and that they were angry.

Because I was also someone who had always been mistreated and angry. That was very, very impressive.

South Korea's ubiquitous militarism is another issue that Jae-sŏk and his friends were mostly troubled by; not only was the presence of US troops in the country a matter of great concern to them, but also the fact that most of the young men in their group would still have to face their mandatory military service in the years to come.[11] Jae-sŏk and several of his friends were playing with the idea of becoming conscientious objectors and serving an 18-month prison-term instead of joining the Republic of Korea Army—a choice that was during those days much discussed among male left-wingers in the anti-US bases movement. Their close friend Hyŏn-jun, I was to learn, had already made up his mind, and was only months away from going to jail when I first met him in 2009.[12]

Hyŏn-jun, then 23, told me that he originally came from Pusan, but he started coming to the Hongdae neighborhood of Seoul already while he was still a high school student living in Korea's second largest city: "It's not like I adored Hongdae, but it was actually the only place in all of Korea where I was able to breathe, that allowed for my cultural survival," he explained. At first, Hyŏn-jun was rather enchanted by the neighborhood because:

I thought of all those small communities in Hongdae as parts of a cultural and political commune movement, and thought it was all really revolutionary. But then I witnessed a lot of those communities falling apart, and then I gave up investing hope into it.

With his new-found identity as conscientious objector and anti-military activist, however, he said that "nowadays I keep wondering for myself if there is anyone here at all who has empathy …"

Hyŏn-jun's exposure to military matters came early, while growing up in a poor neighborhood in Pusan, squeezed in between the US military Camp Hialeah, and the Yangchŏng installation for Pusan's Korean Military Police:

Because the city was so rapidly expanding, these two military bases were all of a sudden located in the middle of the city, so they were scheduled for eviction. What I remember about the military base is the dirty, dusty walls and the barbed wire on top of it. And no entrance signs. That's that—I never saw anybody emerging from the base, and I never got to see the inside of it either. Once a year, there was a really big [display of] fireworks happening on US Independence Day—the entire city was excited about it. Also, it was fun to talk with the other kids about the prostitution district [that was located right next to the base].

Only in 2002, when the death of the two schoolgirls Shim Mi-Sŏn and Shin Hyo-Sun triggered a wave of protests throughout the country, did the proximity of the US military base become something for then 15-year-old Hyŏn-jun to reflect upon. He would soon join the protests, too, and during the rallies, he:

> would swear at the GIs who were sitting in the watchtowers behind the walls, or I would write graffiti on the wall or on the ground, things like "Fuck off, US Army". I am thinking today that my negative perception of all military groups began back then.

A few years after these events, having followed the headlines coming out of the Iraq War closely, he would then join the anti-bases activities at Taechuri, where several other Hongdae punks had also started to get themselves involved.

From Hongdae to Taechuri

Just like Hyŏn-jun, a number of his Hongdae punk friends would be drawn into the left-wing activism surrounding the US military presence in the country during the mid 2000s. One decisive moment that personalized this issue for these young people was the conflict over Taechuri, a small village near P'yŏngt'aek that stood on land slated for the expansion of US Camp Humphreys. In April 2003, the South Korean and US governments had announced their decision to relocate the troops stationed at the Yongsan Garrison in central Seoul to this military base in the P'yŏngt'aek region, which would make an expansion of that post necessary in order to meet the needs of its growing population of US soldiers and their dependents. The Korean Ministry of National Defense then began to contact landowners in the village of Taechuri to inform them that their plots were to be seized, with some compensation money being offered to affected locals. In July 2003, a collection of farmers made up their minds to resist the government's move, with the distinct goal of preventing the Ministry of National Defense from expropriating their farmland (Yeo 2006: 43).

Within less than a year, the local conflict became much larger when a national campaign was founded that brought together over one hundred non-governmental organizations (NGOs), civic groups, and individuals under one temporary framework (Yeo 2006: 44f). Dozens of Seoul activists would now move to Taechuri to support the villagers in their struggle, and, with hundreds of people joining them for days or afternoons whenever they could make the time, the small village next to the US base soon turned into one big bastion of anti-military activism. Every night, a candle-light vigil was held in the elementary school building of Taechuri, with musical performances, poetry readings, and speeches delivered to keep up the spirits of those involved in the struggle; a nightly action that was repeated over 600 times and only came to a halt on 4 May 2006, when the school was finally destroyed in the midst of violent clashes (Yeo 2006: 48).

The coalition initially focused on exploring all legal options and engaging in negotiations with the ministry, but their arguments would fail in the courts. Eventually, matters escalated into physical confrontations that took place in Taechuri itself (Kim J. et al. 2006: 7). On March 15, 2006, for instance, hundreds of farmers and their activist supporters engaged in violent clashes with the riot police that had been deployed to the village in their thousands. And in early May of the same year, 12,000 riot police descended on the village, where they fought with 2,000 activists trying to defend the local elementary school (Yeo 2010, see also Yeo 2006: 34f). Taechuri, at that point in time, looked much more like the DMZ than an ordinary farming community in the center of South Korea; checkpoints, barbed wire, and hordes of young military recruits in full riot gear stationed nearby had become an ordinary sight, making it increasingly clear to the farmers who remained that civilian life in this area was to bow down to the pressures exerted by the local armed forces protecting the interests of the US military.

The intervention of a group of Hongdae punks in this kind of struggle on the very outskirts of the Seoul Metropolitan area was at first viewed as an oddity beyond belief by the elderly farmers in Taechuri, Jae-sŏk and his friend Jil-sung told me. Jil-sung, a 22-year-old punk, explained that there was even "an article written about us on one of those nationalist-leftist newspapers. [T]hey wrote that it was impressive to see these funnily dressed young people guard the village with a bottle of rice wine in their hands [laughs]." People eventually warmed up to the unlikely visitors from downtown Seoul, though, and even asked them to perform their music for them at one of their daily evening gatherings. "We refused ... Jae-sŏk was trying to tell them that some of the old people might be getting a heart attack or something like that if we played ..."

Reflecting on their positionality in the village, especially with regard to the more seasoned activists around, Jil-sung added, "We never thought of ourselves as integrated there. We felt like foreigners there." Much of their sense of estrangement from the activists in Taechuri he attributes to a matter of social distance:

> There were so many student activists there, too. And we felt very distant from them. It was kind of a class issue, I guess. We would think of them as being raised in middle-class families, and slowly getting into reading Marx, and somehow joining Student Unions, while their parents paid for their tuition. Yeah, very different from us.

Jae-sŏk was the one who had initiated their first journey to Taechuri, which the younger Jil-sung attributes to Jae-sŏk having "some kind of charisma. [Back then] you were there and talking about all this new political stuff that we never heard of, and all these kids thought, 'Wow, that's so cool, we should be [in Taechuri], too!'" In order to stay updated with the events unfolding in that village, they would keep themselves informed on a Korean web forum called "Anarclan," where the newest developments concerning Taechuri were being discussed by a small group of netizens with an anarchist background. Jae-sŏk explains how the internet was crucial in how they finally shaped their decision to go to the village to show their support:

> Back then we were all living together, sharing just one computer, so we were all looking at the same stuff [online] together. So we were interested and kept following this issue, and so we got to know the problem better, and we were getting more agitated by it.

After their first visit to Taechuri, they would go there again whenever time, money, and the general mood would allow it. Once, in the summer of 2006, the idea of going to Taechuri came up in the middle of a drinking session at the Hongdae Norit'ŏ:

> *Jil-sung*: So we were all drinking at the Norit'ŏ together ... I was pretty drunk and I suddenly said, let's go to Taechuri. So I said, let's go to Taechuri tomorrow morning. Then we kept on drinking, and everyone was like, yeah, let's go, let's go, that's a good idea. But you and Jae-bong

had your bikes with you. So all of a sudden you announced, "We will go ahead with our bikes now." [...] So you and Jae-bong took off on your bikes in the middle of the night. And then the next day, at around 10am, we called you. "Are you there?" And you guys said that you were sleeping under a tree next to this riot police station [that had been erected near the village].

Jae-sŏk: Yeah, they stopped us from entering the place. And then it was kind of a tense atmosphere. So we couldn't get in.

Jil-sung: Have I told you how I got there that day? I called you guys once I got there, asking where you guys were. I went there the next morning, right after my work was done. I was there, but there was no bus running that was going to Taechuri. So I just stole a bike and rode it there. And the riot police were not checking me at all, they just let me pass by.

The thousands of riot police deployed at Taechuri during those tense months of 2006 were predominantly made up of recent conscripts to the Korean military. Among those young men in uniform that the Hongdae punks faced during the struggle over the village were also a number of friends and acquaintances. Jae-sŏk recalls a strange encounter with an old friend on the outskirts of Taechuri, who was there as part of a riot police unit he was serving in:

I remember that back then on the bus—there was riot police getting on, and then dragging people out of the bus who were not residents of the village. And me and K. and C., we were looking pretty suspicious. But we were hoping that we could perhaps still pass as residents. But then ... one of the riot police who were kicking people out of the bus that day was a guy that I knew from when I was a kid. [...] He didn't recognize me. Or he had just started his service and didn't dare to say anything. Yeah, he was kind of acting like a robot. And I thought that if I talked to him now that it could cause trouble so I didn't say a word.

After the struggle had ended in a defeat for the farmers and activists, Jae-sŏk had an opportunity to talk to some other friends of his who had served there as riot police to discuss their—largely divergent—experiences of the clashes in Taechuri: "Yeah, they were saying something like how they were beating up grandmas [*laughs*]. T., he was in the Special Forces that were destroying the Elementary School. T. says it was pretty horrible." And Jil-sung chirped in: "Yeah, that's how it goes. Just thinking of it—them having to be there—and thinking of all the other students and protesters, and how they could just quit and go home ..."

Not only would the involvement of a number of Hongdae punks in the Taechuri struggle bring about discussions with South Korean military recruits that they knew. It also raised a number of debates among their foreign friends in the Hongdae neighborhood because of the presence of quite a few US military members in the area, who attended Hongdae punk shows or sought to hang out with the punk kids at the playground. In this way, the Taechuri issue was turning into a controversial issue within Hongdae too, as it forced Korean punks and US service members to ponder their alliances and the complexities of belonging to the same alternative scene and sharing the same small entertainment area in their free time, while simultaneously finding themselves on different sides of the trenches when it came to the base expansion issue.

Jil-sung, for instance, acknowledges that in principle there would have been much potential for him to shape friendships with the soldiers he came to meet in Hongdae, but that in practice things got rather complicated with GIs after Taechuri for him. He was struck by the fact that many of them "also didn't finish high school, they were also poor fucks back home. And listened to the same music. But still, in the end it didn't work for me." He relays one instance when he hung out with two US soldiers at the Norit'ŏ that brought this point home for him: "I think I had fun hanging out with these two GIs 'cause they kept buying me drinks, you know. And then I talked to them a few more times ..." Things soon got bothersome for Jil-sung, though:

> I kinda started thinking that is was a mistake to hang out with them after a while, though. They kept annoying me so much, cause they would come to the park, and try to talk to me, and I think it was mainly stuff like, "We want to pick up some chicks, you gotta help us, cause we don't speak Korean."

On another occasion, a GI he knew wanted to discuss anarchism with him, which in Jil-sung's view did not go very well:

> And then I was telling this guy, "You should get the hell out of the military, 'cause it's shit." And he said, "Yeah, you're totally right." He told me he had just bought a punk T-shirt online, and his boss found out about it, and said he couldn't have that inside the base. I think it was related to some kind of political symbols on the T-shirt and he

couldn't have those. And then he was saying something like, "Yeah, I'm an anarchist. I wanna be an anarchist, 'cause my superiors, they are all like ... like commies, you know" [*laughs*]. And I was like, "Yeah, man, yeah, I get it." And ever since then I started to completely ignore him.

Exit the Demilitarized Zone, Enter the Temporary Autonomous Zone?

Despite the many misgivings that Jil-sung and his friends at times voiced towards GIs, it was still striking to see how on a number of occasions I would find them hanging out with GIs like Karl, Mike, and others. Drinks would be shared, jokes exchanged. On occasions, things could get uncomfortable, and one party or the other made sure to move on in time. The common denominator that allowed these Hongdae punks to temporarily sit and party together with GIs can perhaps be attributed to their mutual recognition that they were all socially, politically, and to some degree economically marginalized in the country they found themselves in, and perhaps also to their wish to escape potentially totalizing societies for the length of a night, be it that of late capitalist Korea or that of the United States Armed Forces. The temporary fraternizing, however, was clearly not enough to tear down barriers between people when the hangover subsided the next day.

Paradoxically, Hongdae, even though it is a thoroughly gentrified neighborhood these days, is at the same time also a space that allows its visitors many liberties that cannot be attained in other parts of the country. At times, Hongdae has been the source of much moral panic, as it is understood to be a Korean terrain that is being contaminated by foreigners who no longer stay within the set boundaries of the few spaces that have been allocated to them in this country. While the Korean media have repeatedly focused on the pollution of Korean youth through putatively highly sexualized foreign males, the many other experiments going on in this area, and in particular the small-scale political contestations within Hongdae, have usually been overlooked.

US soldiers, as we have seen, have played a significant role for a while as the putative sources of all evil within this neighborhood. US servicemen, within this context, have also often been blamed for the complex urban processes that have led to the rapid commodification of an "alternative" neighborhood. The anarcho-punks I encountered, however, have mainly criticized them as putatively willing pawns of a militarist-capitalist system that Korea is deeply integrated into as well. The struggle surrounding Taechuri in 2006 was most certainly a key moment for quite a few of these young people—it was their own coming-of-age moment that first allowed them to politically channel and express their at times rather vague feelings of unease amidst hyper-militarized Korea in political terms.

Another point to take note of is the way in which this Hongdae youth is actually diverging from the ideological path laid out by their minjung elders. Politicized by their deep-seated sense that they are misfits in hyper-capitalist Korea, they have learned to engage with the social world around them by looking for ideological clues to be found beyond the borders of the peninsula. Thoroughly disenfranchised from the Korean Dream of rapid development, utterly uninterested in all issues concerning North Korea, and usually countering Korea's putative role as a victim with cynical remarks about their country's increasingly strong involvement in capitalist and militarist projects across the globe, they have to some degree stepped out of the nationalist framework that their minjung elders were so keenly attached to, and have actively looked for inspiration in global radical movements. In such a way, they may have quietly snuck out of the barracks for a while, and opted to live in a temporary autonomous zone (Bey 1991)[13] they have carved out for themselves within the limits of Hongdae.

7

Conclusion

Seeds of Antagonism,
Children of Discord

The legacy of the ville—the sum of the daily economic, social, cultural, and sexual exchanges occurring in camptowns over the decades since the Korean War—still looms large in today's Korea. The long-term stationing of United States forces in South Korea, which has entailed a continuous flow of predominantly young men sent to the country for typically short postings, created an exceptionally pernicious triangle involving foreign soldiers, local women, and native men. Acrimonious sentiments triggered by the issue of camptown prostitution were amplified into a matter that touched upon vital national questions, and have been carried forward through time, affecting ever new generations and urban spaces. In South Korea, then, the aggregation of base encounters involving US military personnel and South Korean women has over time widely eroded the sense that the entertainment spaces near US military installations are inevitable or natural outcomes of the presence of US forces in the country. All the while, it has been in the face-to-face, everyday encounter with soldiers who work in these bases that the highly contentious spaces adjacent to military installations *did* become normalized for many individuals who experienced them, with novel configurations of gender, power, nation, and class inexorably emanating from these areas into the rest of the country.

While the South Korean case presented here has fundamentally been shaped by a series of historical, political, and economic contingencies, some aspects of this story can arguably be generalized. That is, resonances and similarities in other locations that have also been drawn into the US empire of bases should not be overlooked. Linda Angst, for instance, has shown how the public contestations that erupted after a young schoolgirl was gang-raped by three US servicemen in Okinawa swiftly moved away from a feminist focus toward a nationalist one (Angst 1995). And in the aftermath of the murder of Philippine transwoman Jennifer Laude, for

Figure 7.1   Connecting places, changing speeds

which a US sailor was convicted of manslaughter in December 2015, the issue of rampant violence against members of transgender communities was equally quickly sidelined to make space for countless discussions over the vulnerable state of the Philippines' sovereignty today.

To be sure, for as long as troops have moved about in other people's countries, the presence of camp followers, and fractious relations between soldiers and civilians, have persisted, causing anxiety and resentment among local populations. Violent imaginaries as I have described them here, however—referring to the social practice of making sense of US militarism through the reconfiguration of individual acts of violence into an issue that pertains to the nation—are a form of engagement with foreign troops that is of a much more politicized order. What is more, these imaginaries, as we have seen, have become deeply embedded in and imbued with the particularities of the spaces they have emerged from or gradually came to touch upon. Violent imaginaries, then, in the way they are being utilized by various actors today, often get entangled in and are complicated by place-making projects, the construction of local territories, and cultural contestations that occur near US bases.

Perhaps it is because violent imaginaries are so malleable in the way they interact with micro-histories and specific contexts that they can arguably also function as a rather flexible transnational formula. Younger, internationally well-connected anti-base activists, in particular, have used these imaginaries in order to constitute a form of meaning-making that travels well to other places across the globe. South Korean activists whom I met in Seoul not only wanted to talk to me about crimes committed in Tongduch'ŏn, but would also often invoke Subic Bay, Okinawa, or other key locations in the global empire of US bases that they were familiar with as well. The transnational conversations these actors are involved in (who have often encountered Filipino, Japanese, Hawaiian, or other activists at conferences, workshops, and protests) have certainly contributed to a pro-liferation of violent imaginaries as a popular form of contestation across the world. At the very least, Korean violent imaginaries have, through such disseminations, absorbed new histories and specificities, and have, in turn, fed into already existing anxieties over sexual and other forms of violence committed by US soldiers at other key nodes within the chain of bases that the US has spun around the globe. In this way, violent imaginaries in South Korea are just one example of a number of conflicts triggered by base encounters taking place worldwide, with social disruptions caused by the overseas presence of the US Armed Forces repeatedly being channeled into this particularly potent form.

Korean contestations over US bases in the country, however, are unique in that the agreement between the USA and the Republic of Korea (ROK) has a long history, is seemingly open ended, and directly involves a number of urban terrains that are (in)conveniently located within reach of US military posts. While the current stationing of approximately 30,000 US soldiers may also reflect new American strategic thinking toward the rise of China, which has become a crucial factor in the US maintaining such a large military presence in Asia and the Pacific, at the same time US troops in South Korea are also a mark of the continuing unresolved tensions between the two Koreas. With the "division system" (Paik 2009) that engulfs the Korean peninsula representing one of the final vestiges of the Cold War, historical particularities in this region have engendered two hostile, yet strangely complementary state arrangements: a kind of "socialism of the barracks" in the North, and a "capitalism of the barracks" in the South. These (in)compatible regimes, interestingly enough, would both opt to model entire social sectors after military structures as time went on; a move which was regularly defended by local officials by pointing at real or imagined external enemies waiting at the gates.

In South Korea, however, these local forms of militarism (born out of a succession of autocratic regimes that could count on being supported by Washington) have never received as much attention as the US Armed Forces presence in the country did. Nationalist-driven hostilities over US bases in Korea, as we have seen, gained momentum in the late 1980s and early 1990s. The seeds of antipathy, however, had been sown decades before the heinous Yun murder of 1992 made national headlines. Following World War II, the Americans had invested many resources and much manpower into maintaining the Republic of Korea, even if that often meant supporting anti-democratic forces within the country. All the while, stark economic asymmetries between the USA and the ROK expressed themselves in high dependency on American assistance in general, and on the US dollar and those individuals who brought it into the country in particular.

Amidst a US *Realpolitik* that involved backing repressive violent regimes in South Korea at the cost of democratization, US soldiers were not exactly model ambassadors on the ground, which to many Koreans increased the unease caused by their country's long-term subordination to US interests. To the contrary, these young male Americans, socialized into a military culture that frequently nourished a virulent form of hyper-masculinity, would often engage in petty crimes and prostitution around US military bases, with the occasional rape and murder until the 1990s going unpunished in Korean courts, thereby increasingly infuriating ever growing sections of society. Male Korean writers of a dissident background, in particular, took grave offense; and by expressing their rage in the form of a "camptown literature" (Lee J. 2012), they laid the foundations for a counter-hegemonic practice that was the first to represent US soldiers as foreign villains ruthlessly pursuing local women in the entertainment areas nearby US bases.

Over the last few decades, such resentments held against GIs have become more systematically utilized by nationalist left-wing forces in the South. In their attempts to reshape USA–ROK relations, violent imaginaries concerning US servicemen have played a central role, with this particular social practice amplifying violent acts committed by US soldiers into epitomes of the uneven relationship between the United States and this small East Asian nation. And while US soldiers were in this way increasingly positioned within a long historical line of intruders that have repeatedly violated Korea's sovereignty, its terrain, and its women, public contention over various GI crimes has led to a widespread change in attitudes among the Korean population toward US soldiers stationed in their country.

The murder of a young Korean sex worker in the early 1990s, we have seen, led to a kind of "structural amplification" (Sahlins 2005) that proved to be essential in popularizing the image of US soldiers as violent brutes on the loose in the adult entertainment spaces nearby American bases. The image of mutilated Yun Kŭm-i, which was spread across the nation with surprising speed, condensed a number of long-held grievances, thereby allowing the conjuring up of support for the growing anti-base movement. Evoking familial and sexual anxieties among (male) citizens of the nation, the actual force of such a bodily image of the nation as a ravaged woman, it seems, can be found in the fact that it is indeed a metaphor that is very easy to think and act with during times of turmoil. The picture of a murdered woman thus became as a key component in the social practice of violent imaginaries, as the horrors depicted paradoxically allowed the visualization of a previously unthinkable form of grassroots politics against the current state of the ROK–US military alliance. After the Yun murder of 1992, then, violent imaginaries were increasingly turned into a near-mainstream frame that was utilized by many to re-evaluate the presence of US soldiers in the country, with the entertainment spaces in which servicemen spent their free time being re-cast as spaces of exceptional violence, too.

The much discussed "camptown problem" (*kijich'on munjae*) in the end also marshaled enough outrage that the economic side of the business with women was affected, as club owners from the early 1990s onwards found they had trouble maintaining a steady supply of females into kijich'on. Eventually, this issue was solved by camptown entrepreneurs through the introduction of foreign entertainers hailing from the Philippines and the former Soviet Union, with local anger over the degradation of Korean women now giving way to international concerns over the putative sex trafficking of foreign women into Korea. While the ways and means through which these women arrived in South Korea have certainly often been illicit and at times outright deceptive, the foreign women, once they came to South Korea's camptowns, frequently tried to better their lot by putting all their hopes and desires into finding themselves a "nice GI" to get married to. But as far as I could tell, their "preoccupations" (Hewamanne 2013) with US servicemen more often than not turned out to be somewhat anxiety-ridden and fragile future-making projects as, increasingly, their clients would opt to escape the "ville" in their free time and go to central Seoul instead, where sexual encounters could more easily be had for free.

In this way, these deeply asymmetrical terrains nearby US bases, and the affective "preoccupations" these spaces have produced, have been reconfigured into a new transnational form, thereby prolonging a troubling inheritance that most ordinary Americans are utterly unaware of, but that their soldiers have to grapple with on a daily basis during their deployments on the Korean peninsula. And indeed, in South Korea the consternation over the very existence of these entertainment areas has never entirely dissipated; it lingers on and is ready to be re-activated whenever "critical events" (Das 1997) involving US soldiers serve as vivid reminders to the local population of the persistence of US military camptowns in their country.

The day-to-day experiences of various actors brought about by the highly charged rendezvous between civilian downtown entertainment spaces and the US military has been the other main focus of this book. Sherry B. Ortner has once noted that, regardless of our focus on the intentions, motifs, and aspirations of the actors we follow around, actual social change at our field sites may often turn out to be "largely a by-product, an *un*intended consequence of action, however rational action may have been" (1984: 157). I take this remark as a solid reminder of the empirical fact that larger-scale contestations over US bases in South Korea, in spite of their clear political directionalities and the at times massive popular support they conjured up, have not (yet) attained the one goal that was central to all these efforts; that is, the removal of US posts from the country. At the same time, we have seen that these political struggles have produced myriad unintended consequences whenever they "hit the ground" in a particular camptown, neighborhood, or entertainment district. And during field research in and near Seoul, I could indeed witness a number of surprising adjustments, minor permutations, and small-scale corrosions affecting the rather powerful bigger picture that left-wing actors have painted of US servicemen in the country.

For instance, although It'aewŏn, the entertainment district in central Seoul that is most closely affiliated with the US military, often gave rise to widespread fear and concern over the presence of large numbers of putatively violent foreign soldiers in the area, at the same time this neighborhood turned out to be a site featuring a disparate cast of actors, where many unusual connections were forged between seemingly incommensurate audiences, and some rather creative appropriations of the negative images of camptowns took place. Here, in Seoul's inner-city interstitial space between the USA and South Korea, I found that many of the ambiguities and uncertainties of violent imaginaries were worked out in often unique ways. This particular social practice, the origins of which I explored in the first three chapters of this book, was in this neighborhood often captured, undermined, and reconfigured amid the manifold everyday encounters between soldiers and civilians.

Through such open-ended meetings taking place in the entertainment areas in and near Seoul, in turn, urban realities were constituted that produced and to some degree institutionalized unlikely spaces for exchange, dialogue, and contestation between a vast set of actors. I found that, contrary to widespread discourses focusing *only* on violence and exploitation, the everyday realities that I witnessed in some instances gave room to the emergence of an unlikely sense of *communitas* between the GIs and male and female civilian inhabitants or visitors to these areas of Seoul. Victor Turner, in his famous invocation of the term "communitas," has sought to describe the comradeship that may arise in temporary realms outside of ordinary social structures (that is, in moments "in and out of time"), where the everyday ties that organize society are renegotiated, (con-)tested, and eventually re-affirmed (see Turner 1969: 96). In a somewhat similar fashion, entertainment areas in Seoul can be viewed as spatial manifestations that allow both civilian and military participants to "let off steam" via the establishment of a strange and typically short-lived camaraderie, which allows everyone to return to the hardships of their respective everyday lives with renewed verve. From the dark corners of It'aewŏn's Hooker Hill to the well-lit territories of Homo Hill, from the noisy realms of the Hongdae playground to the tightly packed dance clubs of said area—Seoul's entertainment spaces occasionally surprised me with their potential to enable different modalities of engagement between locals and young soldiers.

The highly volatile socialities made possible in these urban spaces at times also worked against the logics and directionalities of both the US Armed Forces and the South Korean state. Both institutions repeatedly sought to enforce a strategy of containment: keeping soldiers separated from locals as well as keeping them out of urban areas was, until quite recently, seemingly the most significant way of preventing perilous encounters and the dissemination of dangerous ideas among servicemen and civilians. From the threat posed by putatively STD-ridden Korean sex workers that US military officials sought to counter through cooperation with local forces during previous times, to the potential dangers of long-haired Korean rock musicians "contaminated" by their prolonged exposure to Western culture that Park Chung-hee feared, to the corrosion of young minds and bodies in Hongdae that have seemingly been ruined by foreign influences—kijich'on spaces and inner-city entertainment areas that are popular with GIs have often been viewed as fundamentally endangering the morals of those who are drawn into them.

Such strategies of containment, it is safe to say, have failed rather spectacularly. Not only have the soldiers found ways to temporarily escape the areas that have been assigned to them, with Hongdae, in particular, emerging as the new favorite space of many GIs; young Koreans and civilian foreigners, too, have begun to invade the former camptown of It'aewŏn *en masse* since the days of the dictatorship. And, in a curious fashion, the role that Hongdae has started to play for the wider area of Seoul over recent years indeed resembles the avant-garde function that It'aewŏn held in those decades after the Korean War. Hongdae, much like It'aewŏn, is a radically liberal(izing) space of consumption, a terrain that allows for the quasi-carnivalesque inversion of ordinary and orderly Korean and US military life, where for the length of a night, anything and anyone can be experimented with. Consequently, short-lived sexual encounters and ephemeral fraternization among the strangest assortment of civilian and military actors are seemingly the order of the day in both of these spaces.

The outrage caused by the "contamination" of the entertainment area of Hongdae, as we have seen, has proven to be infinitely greater than controversies over It'aewŏn, though. Perhaps this has to do with how It'aewŏn has been considered an extra-territorial space for such a long time now; a terrain that in the eyes of many had to be sacrificed in order to keep GIs contained in one place. Be that as it may, the public image of Hongdae is vastly different from that of It'aewŏn in the way that Hongdae is for the most part understood to be Korean terrain, a space that is nowadays slowly but surely being infiltrated by foreign forces, which makes the moral affront that this neighborhood represents to both a conservative and a left-leaning nationalist Korean public all the more visceral.

To be sure, the rapid change that South Korea has undergone over the course of a lifetime has bred much unease in a country where the main preoccupation of both left- and right-wing forces during much of the 20th century has been to come to terms with issues related to imperialism, self-determination, and the constant threat of war. The central motif of a "nation under siege," when deployed with regard to highly contested spaces like Hondgae, thus brings out perennial fears of clandestine infiltration and corruption of the nation's core values. Therefore, much consternation in the media and in the public sphere could be seen over issues pertaining to jovial fraternizing and sexual relations formed between locals and foreigners, with Hongdae coming into the spotlight over the last decade for providing the very space for these putatively poisonous encounters. In this explosive context, contestations that used to be firmly located within camptown areas, such as those over the figure of the "yanggongju," were now revived in an inner-city entertainment area that is seemingly worlds apart from the seedy corners of kijich'ŏn.

All the while, Hongdae, much like It'aewŏn, too, is a multifarious space, a terrain that can in principle accommodate many versions and visions of what the neighborhood is all about. In reality, however, in both areas we could witness a number of ongoing place-making struggles; that is, contestations over who gets to call the shots and gets to inscribe their own visions into the urban regions of Seoul. Within such a heated atmosphere, various actors have incorporated elements of violent imaginaries into their strategies of territorialization. In this way, in inner-city Seoul, a free-floating discourse that has sailed through the political and media landscape of South Korea whenever another "GI crime" occurred, was turned into a deeply territorial practice that ultimately became about inscribing social, political, and cultural projects onto various neighborhoods. GIs and other rowdy male foreigners, as we have seen, became easy targets amid contestations over Hongdae's moral ownership; they simply represented the type of people that a number of local actors could readily agree upon should be kept out of the neighborhood.

The leftist punks that we encountered in the last chapter, and who were branded as prime examples of Hongdae's thoroughly corrupted youth by the media, are very much taking part in such place-making games. They are, as we have seen, exceptional in that they are strongly concerned with how to circumvent, contest, and subvert *both* home-grown *and* foreign forms of militarism. Looking deeply, and at times rather disapprovingly, at the South Korea of today, they see a vastly changed country which has made the stellar climb from the margin of the world economy into its very heart, but still has many legacies of authoritarianism and militarism to tackle, including those brought about by their own elites.

In their particular involvement with matters of autonomy, they may at first sight resemble the minjung activists of the previous generation, who were also troubled by questions that had to do with sovereignty. Self-government, however, is framed rather differently by these factions—while minjung actors sought to change the *kind* of state they found themselves in, these young anarcho-punks would in principle like to see the power of the state they live in erased altogether. In a good anarchist fashion, in the meantime they tried to settle for the creation of what the anarchist poet Hakim Bey (1991) has called "temporary autonomous zones." This kind of political vision is also what has made the struggles surrounding the village of Taechuri so interesting to them: the short-term declaration of independence of a handful of aging farmers inspired these young men and women to make their way from Hongdae to a little hamlet a few hours outside of Seoul, where they joined hundreds of other activists who were dreaming of alternative ways to organize life that would not involve military conventions. Unsurprisingly, the dream did not last, but the experiment itself is still much talked about among those who witnessed or had heard about it, with the image of the defiant Korean village for a little while making its way into news and activist channels across the globe (e.g. BBC 2007).

While South Korean contestations over US bases have centrally been about the perceived state that this small East Asian nation finds itself in today, at the same time they also form a part of the wider, transnational movement against US military installations overseas. Okinawa, the Philippines, Diego Garcia, Guam, or Hawai'i—in the last analysis, none of the small-scale conflicts over US bases that have occurred in various locations around the world can fully be grasped without keeping the large-scale economic and political shifts within the global capitalist order in mind that have only accelerated during the last decade. South Korea, as I have attempted to show in *Base Encounters*, is an excellent case in point, as its particularly turbulent history and its economic transformations during the second half of the 20th century have given rise to an acute and widespread sense among the population that the country has at times been treated rather unjustly by its American ally. The new confidence with which such declarations are made is inarguably founded in the spectacular economic and political successes of the country—a confidence which is likely to only rise, given the current turbulence affecting the "Global North." While South Korea, positively influenced by the re-emerging powerhouse of China, has recovered at astonishing speed from the economic crisis of 2007/08, the United States, the European Union, and many of the other nation-states that together make up the core countries of the world economy, are still stuck in a sense of decline that was simply unimaginable just 20 years ago. Shaken to the very roots by the global economic turmoil that has caught these countries by surprise, discouraged by the seeming futility of the drawn-out military engagement in the Middle East that has only given rise to more upheaval in the region, and increasingly challenged by both nationalist uprisings and far left movements on their own turf, these core countries of the capitalist world are seemingly plunging ever more deeply into a state of crisis. Seen in this light, the South Korean case that I have presented here is also the story of on-the-ground, everyday changes that mirror, complicate, and feed back into the tectonic power shifts we are witnessing these days.

South Korean changes in attitude toward the representatives of the most powerful military in the world, I have argued, can only be grasped in their multi-dimensionality by taking both a close look at vast transformations in the national, regional, and global order, and at how such changes reverberate in the mundane zones of contact between different actors from the West and the rest. The particular challenges ahead for relations between the US and South Korea, and the future of US troops stationed in the country, are hard to predict. What can be stated with relative certainty, however, is that the days of unassailable US supremacy in the region are over. The violent legacies, perilous imaginaries, and deeply ambivalent encounters that the US military presence on Korean terrain has given rise to, I also dare to prognosticate, will haunt all parties involved for many more years to come.

Notes

1 Introduction: Violent Imaginaries and Base Encounters in Seoul

1. This book is based on ethnographic field research in and near Seoul, South Korea, which I conducted from September 2007 until June 2009. During the first year of my stay in Seoul, I enrolled in a sequence of five intensive language classes. While the Korean women who worked in the bars and clubs nearby US bases typically preferred to speak in English with me (as did the Filipina entertainers, and many of the other actors I met in these entertainment areas who considered these districts as "extraterritorial" spaces), my gradually increasing knowledge of Korean most certainly proved crucial in my interactions with NGO personnel and activists. Much of the information given in this book was gained through participant observation and informal conversations with a number of actors I encountered in entertainment areas nearby US bases or in Hongdae, with discussions typically taking place over drinks or coffee in a noisy bar or club. While Hongdae and It'aewon were two neighborhoods in which I could easily (and for the most part without much risk involved) meet people on my own, gradually "snowballing" my way from one informant to the next, access to informants in the more remote and secretive kiji'chon areas outside of Seoul was made possible through introductions provided by staff of the counselling center of Turebang. In all the (semi-)urban entertainment spaces I explored for this book, however, only a few people (typically activists) felt comfortable taking part in formal interviews. US soldiers and Filipina entertainers, in particular, feared that there would be negative repercussions if it was to become known that they had given an interview to a researcher. In order to alleviate my informants' concerns, I usually refrained from coming to meetings with a predetermined set of questions, or from taping our conversations. In a few instances, too (for example, when visiting night clubs that employ Filipina entertainers in the company of Turebang staff during "outreach" events) full disclosure of my role as researcher proved to be impossible, as it could have potentially led to dangerous situations for myself, or those around me.

2. Believed to originally have been an abbreviation for "galvanized iron," this colloquial term is often used nowadays to describe members of the US Army. In South Korea, it is a popular word deployed by the local population as a name for any US Armed Forces personnel, regardless of the actual branch they may belong to. In the following, I shall also use GI as a generic replacement for "US soldier" or "serviceman."

3. The number of actual "boots-on-the-ground" is a matter of much speculation. Following a 2008 agreement between Bush and Lee Myung-bak, his counterpart at that time, the USFK for a while maintained its strength at 28,500 soldiers. However, following recent tensions between North and South, and as a result of current president Obama's "pivot to Asia," rotational deployments to South Korea have increased in frequency. The number of additional soldiers is quite significant: in September 2013, for instance, a 9-month deployment of 380 soldiers of the 4th Squadron, 6th Cavalry Regiment from Joint Base Lewis-McChord to South Korea was announced, and, in January 2014, another 800 troops of the 1st Battalion, 12th Cavalry Regiment were brought in temporarily. While these additional troops are not included in the official numbers provided by the USFK, they suggest that, as of April 2014, the number of US soldiers in South Korea stands at at least 29,680 (Capaccio and Gaouette 2014; Rowland 2014). Korean sources have claimed that the number of troops in the country is even higher. Citing the US Department of Defense, *Hankyoreh* has reported an increase of over 9,000 soldiers, with 37,354 soldiers in South Korea as of September 30, 2011 (Kim K. and Gil 2013).

4. While the percentage of women in active duty in the US Armed Forces has grown from 2 percent in 1973 to 14 percent in 2010 (Patten and Parker 2011), this book will focus on imaginaries and encounters that have male soldiers at their center. The positionality of female US soldiers in South Korea, while an extremely interesting topic in its own right, will not be addressed.

5. All names of individuals (and, if deemed necessary, other locational or occupational details that could jeopardize anonymity) have been changed according to the usual anthropological standards of privacy protection. For further information on ethics in anthropological fieldwork, see, for instance, http://www.aaanet.org/issues/policy-advocacy/code-of-ethics.cfm.

6. *Kimchi* is a popular Korean side-dish that consists of spicy, fermented cabbage.

7. A crucial problem when trying to write about the women I encountered in kijich'on had to do with choosing a name for their work, as these migrant women tended to go by many labels in South Korea: they themselves often chose to call themselves "entertainers," a very utilitarian approach perhaps, as it simply refers to their visa status in the country, with the vast majority of them having been holders of E6-entertainment visas. Their customers, too, were usually not labeled as such, but women usually referred to them as "boyfriends." "Juicy girl," or simply "Juicies," on the other hand, was the name that the clients, the soldiers, gave them, a double reference to the girls being "juicy"—that is, sexy—and also referring to the fact that they constantly drink overpriced fruit juice drinks that the soldiers have to buy them if they want to spend time with an entertainer at a club. Finally, at the non-governmental organizations (NGOs) I encountered, they were usually referred to

as "prostituted women" or victims of trafficking, while to the Korean public, they were often just understood to be "bad women."

8. All references are to US dollars.

9. We can only begin to appreciate the fundamental role the US has played for the South Korean defense apparatus when we take into consideration that, right up to today, OPCOM (operational command)—that is, the command over both South Korean and American troops in case of war—is still in the hands of the USFK. Only since 1994 has the USFK handed over peacetime command of the ROK Armed Forces to the Korean side. On the difference between operational command and actual control in the South Korean–US military alliance, see Drennan (2005: 291f).

10. Anthropologist David Price, in his historical work, has documented in great detail how the US military and various US secret services have time and again collaborated with a number of renowned anthropologists. On the subject matter, see his two books *Anthropological Intelligence* (2008) and *Weaponizing Anthropology* (2011).

11. It is important to note, for instance, that normative attitudes against the US Armed Forces presence among my Korean informants do not necessarily entail a more general anti-militarist stance. The opposition to *homegrown* forms of militarism, as we shall see, is a much smaller force in South Korea than the opposition against US bases in the country. Historically, the US military may have even been made to take the blame for things that South Korean generals did at times, thereby underscoring the point made by Martin Shaw that the emergence of anti-militarist movements is a highly contingent process that is mediated by a number of socio-economic, political, and ideological forces (2012: 31f). Given the omnipresent power of South Korean military institutions in the country to this day, local manifestations of militarism have proven to be much more difficult to approach than those of the US military, with proponents of the Korean nationalist left generally steering clear of discussing the local armed forces in order to avoid alienating large sections of society, who generally support the ROK military. Perhaps, then, the US military has at times also had the function of being a lightning rod for broader issues of militarism on the Korean peninsula, which local activists still find incredibly hard to approach directly.

12. In February 2000, 120 liters of formaldehyde were disposed of by pouring it down a drain inside the Yongsan US Army garrison. In this way, a civilian mortician was seeking to get rid of the formaldehyde, used as an embalming fluid inside a mortuary at the US base in central Seoul, with the substance finally ending up in the Han river (that is, one of the main water sources for residents in the capital area [Scofield 2004]).

13. For more information, see the website of the National Campaign for Eradication of Crimes by U.S. Troops in Korea (http://usacrime.or.kr/doku/doku.php).

14. With "traditional Chicago School or central place models of concentric land-use patterns surrounding centralized metropolitan cores" having become a rather outdated perspective for the understanding of the majority of cities in the world today, cities such as Seoul are perhaps better understood as "massive, polycentric urban regions" (Brenner 1999: 437). Friedmann and Miller's concept of the urban field (1965), too, might be successfully applied to the way the city of Seoul interacts with its hinterland these days:

> The urban field may be viewed as an enlargement of the space for urban living that extends far beyond the boundaries of existing metropolitan areas—defined primarily in terms of commuting to a central city of "metropolitan" size—into the open landscape of the periphery. (1965: 313)

15. Official census for 2015, see: http://www.index.go.kr/potal/stts/idxMain/selectPoSttsIdxSearch.do?idx_cd=2911&stts_cd=291102&clas_div=C&idx_sys_cd=540&idx_clas_cd=1

16. In the field of social movement studies, there exists a wide body of literature on "framing." The notion was first introduced into the debate through Erving Goffman's (1974) book *Frame Analysis: An Essay on the Organization of Experience*. Social movements, Robert Benford and David Snow later contended, are not merely "carriers of extant ideas and meanings that grow automatically out of structural arrangements, unanticipated events, or existing ideologies." Rather, they argue for a focus on the "movement actors" who are, in fact, "signifying agents actively engaged in the production and maintenance of meaning for constituents, antagonists, and bystanders or observers" (2000: 613). For our case, "injustice frames," in particular, are of interest—that is, the way in which social movements identify the putative victims of a given unjust situation, and establish who the culprit is as well (2000: 615). An understanding of knowledge production within social movements along the lines of Benford and Snow's argumentation thus implies a *dynamic*, *deliberate*, and heavily *negotiated* production of frames by key actors.

17. Using the term "social practice" is a nod toward Sherry Ortner's influential article "Theory in Anthropology since the Sixties," in which she answers the question of what a social practice may be in the following way:

> In principle, the answer to this question is almost unlimited: whatever people do. Given the centrality of domination in the model, however, the most significant forms of practice are those with intentional or unintentional political implications. Then again, almost anything people do has such implications. So the study of practice is after all the study of all forms of human action, but from a particular—political—angle. (1984: 149)

18. The anthropological body of literature on violence is, as Kamala Visweswaran has pointed out, "largely an assemblage of ethnographies of political violence from around the globe, comprising a range of forms—ethnic violence, genocide, dirty wars, revolutionary violence, peacetime crime, gendered violence, torture, militarization—and the themes of fear, trauma, and memory work" (2014: 19). Soldiers are experts in violence, with much of their military training revolving around the issue of how to exert force in the most effective manner; and they have frequently been singled out as the central protagonists in this sub-domain of social anthropology (see, for instance, Frese and Harrell 2003; Grassiani 2013; Gutmann and Lutz 2010).

2   Capitalism of the Barracks: Korea's Long March to the 21st Century

1. The degree of modernization undertaken by local elites before Japanese imperialist encroachment is a matter of heated debate both within Korea itself and among foreign scholars working on the subject (Cumings 2003: 282; Eckert 2000: 6ff).

2. Following the destruction of a trading ship (the *General Sherman*) in P'yŏngyang in 1866, the United States conducted their first military action on Korean territory in 1871 by sending five warships on a punitive mission into Korean waters.

3. After the killing of nine French missionaries and several thousand Korean Catholics in 1866, France undertook several retaliatory actions against Korea.

4. Russia's imperial postures toward the Korean peninsula were a major cause of the Russo-Japanese War (1904–5).

5. Empress Myŏngsŏng, also known as Queen Min, had attempted to undermine Japan's increased influence on Korea by signing treaties with Russia. On October 8, 1895, she was murdered by several assassins who had infiltrated the royal palace and were said to have been hired by the Japanese Minister to Korea, Miura Gorō.

6. "Social Darwinism," explains Korean studies expert Vladimir Tikhonov, "already well-established in Japanese Meiji discourse on world and 'nation', was introduced in the role of an all-embracing paradigm, cosmic and social, cognitive and ontological [narrative]—a role it hardly ever played in its Western 'homeland'" (2003: 82).

7. After Japan's victory over Russia during the Russo-Japanese War (1904–5), Japan forced Korea to sign a treaty that turned the peninsula into Japan's colonial protectorate, thereby paving the way for a full annexation of Korea that was to take place in 1910.

8. For an excellent article explaining the intricacies of Japanese ethnic nationalism and the way it sought to incorporate the nations it saw itself surrounded by, see Doak (2008).

9. Numerous books and articles have been published over the last two decades on the so-called comfort women issue—English publications include, for instance, Chai (1993), Hein (1999), Hicks (1995), Soh (1996, 2009), Tanaka (2001), Yang (1998), Yoshimi (2002).

10. On US military rule, see *Relations between US Forces and the Population of South Korea. 1945–2010* (2014), written by French historian Bertrand M. Roehner, who is also the author of *Relations between Allied Forces and the Population of Japan, 15 August 1945—31 December 1960* (2007).

11. "The class dimension of the [1961] coup," writes John Lie, "cannot be ignored. Many military officers [involved in the coup] hailed from humble social backgrounds" (1998: 47).

12. Nothing shows Korea's economic successes in a more straightforward manner than the development of the country's per capita GDP: from $82 in 1961, it climbed to $318 in 1972, and reached $2,588 in 1980, the year after the end of Park's era (Adesnik and Kim 2008: 7).

13. This link between contracts doled out by the US military and the rise of a Korean conglomerate is by no means unique to Hyundai. Rather, it seems to be an innate, but understudied feature of the spectacular rise of these South Korean enterprises over recent decades (Glassman and Choi 2014).

14.

> In three- and four-story warehouses small garment manufacturers would create platforms about four feet high, and in every available space put a table, a sewing machine, and a young woman. Dust, dirt, heat, and cotton particles would blow through this small space, which has no proper ventilation. Ten to twenty young girls, unable to stand upright, would trundle in to squat in front of the whirring machines. Put a thousand of these shops together and you have the Peace Market, employing about 20,000 workers in all. (Cumings 1997: 374)

15. The KCIA had also become a lethal weapon of terror over the years. Among countless incidents, particularly noteworthy here are the kidnapping and near-killing of oppositional leader Kim Dae-jung in 1973, and the abduction of 17 Korean nationals from Western German soil who were later put on trial in Seoul facing fabricated accusations of having founded a spy ring in Berlin (Kim M. and Yang 2010). For an account of the life of Yun Yisang, the left-wing composer at the center of the so-called East Berlin Spy Incident, see Rinser and Yun (1977).

16. This refers to people who were in their 30s at the time this term was invented (i.e. in the 1990s), who had gone to college in the 1980s and were born in the 1960s (Park M. 2005: 267).

17. On the continuing controversy over the degree of US (military) involvement in the Kwangju massacre, see for instance Drennan (2005), Katsiaficas (2006), Shorrock (1999), Wickham (2000).

18. This invitation was in fact meant as a reward for Chun Doo-hwan in exchange for not having oppositional leader and Kwangju native Kim Dae-jung executed, but rather, "only" sentenced to a life of imprisonment (Cumings 1997: 383).

19. Kim Dae-jung, in contrast to the military men before him, was driven by a new sense of nationhood and masculinity, argues Sheila Jager, with the idea of struggle for Kim taking on "an entirely different meaning than it had in the past, becoming much more allied with the Christian idea of suffering and redemption than with social Darwinian notion of progress" (2003: 143).

20. From 1998 onwards, South Korean tourists were allowed to visit the Kŭmgangsang area, a mountainous region in the eastern part of North Korea, with hundreds of thousands of South Koreans having made use of this opportunity up to July 11, 2008. On that day, the death of a South Korean woman, shot dead by North Korean military personnel while going for a stroll on a beach and accidentally crossing a military line, would bring a halt to the tourist program, which is not likely to be continued any time soon given the currently hostile climate on the peninsula.

21. Kaesŏng, a North Korean city that is relatively close to the DMZ, has become home to an Industrial Zone that houses over one hundred factories built by South Korean companies since 2002, in which tens of thousands of North Korean workers produce goods for a fraction of the wages South Korean workers would need to be paid. During the spring 2013 crisis, all economic activity at Kaesŏng was shut down (and to date has not been rekindled).

22. On May 17, 2007, a train crossed the border between North and South Korea for the first time since the Korean War—this widely reported test run, however, did not lead to regular services starting up.

23. On the emergence of South Korea's vibrant NGO scene in the late 1980s and early 1990s, see for instance Isozaki (2002), H.R. Kim (2003), H.R. Kim and McNeal (2007), H.K. Lee (1996).

24. Alexander Cooley (2005) writes that in "June 2002, usacrime.org.kr also posted several photographs of the dead bodies of the two schoolgirls from the Highway 56 incident, several of which flooded hundreds of thousands of inboxes just hours after the photographs were taken."

25. In a scathing critique of the Lee Myung-bak administration's "rollback on democracy," sociologist George Katsiaficas (2009) notes that during the emotional days following former President Roh Moo-hyun's suicide (in the midst of bribery allegations brought against his family by a committee closely associated with the new government):

> police buses encircled a memorial site in Seoul for former president Roh, and riot squads refused to open their cordon of buses, compelling thousands

of people bringing incense and prayers to line up through subway stations. Nearly 1,000 police were deployed in front of the memorial at Deoksugung Palace; altogether over 8,000 police were sent into the streets for crowd control.

26. In a 2002 memorandum on Korea, Donald Rumsfeld, who was then US State Secretary of Defense, noted that: "We do need to rearrange our footprint there. We are irritating the Korean people. What we need to do is have a smaller footprint, fewer people, and have them arranged not so much in populated areas" (Ramstad 2011).

3   "The Colonized Bodies of Our Women …": Camptown Spaces as Vital Zones of the National Imagination

1. Detailed information about the murder (both in Korean and English) can be found on the website of the National Campaign for Eradication of Crimes by US Troops in Korea (http://usacrime.or.kr/), an organization that was founded in the wake of the Yun murder and actively monitors US military-related crimes up to this day. A 2010 article in English by blogger Robert Neff also provides a good summary of the actual crime, even though key points raised by the author are highly speculative.

2. The subject matter could also be approached in light of a recent "pictorial turn" in the humanities and social sciences that was to a large degree inspired by the work of W.J.T. Mitchell. His groundbreaking book *What Do Pictures Want?* (2005) has recently also inspired a number of anthropologists such as Birgit Meyer (e.g. 2011) or Christopher Pinney (2011), who focus on the role of visual culture for processes of mediation and imagination in religious settings.

3. The significance of this event for Turebang and its struggle in the camptowns can be seen in how much space the group itself allocates for this event in its self-introduction on its website (www.durebang.org): "People became aware of the seriousness of the crimes of American soldiers and the unfair agreement over how to handle these problems made between the governments of Korea and America," it says about the event, and further explicates: "After the murder case of Yoon, centers for the reporting of crimes by American soldiers were created in different areas of the country."

4. Such attempts to draw similarities between the "comfort women" system and the ongoing prostitution in camptowns, or perhaps to even forge a joint movement, have proven to be difficult (Moon 1999), with some of the surviving victims of the "comfort women" system, in particular, objecting to the comparison. My first encounter with this problem came during a visit at the "House of Sharing," a home for some of the survivors, where we were

shown a documentary in which one former comfort woman expressed very strong opinions about how they had nothing to do with common prostitutes like the ones laboring in camptowns. A volunteer, who lived together with the women at the House of Sharing also confirmed in a conversation with me that several of the survivors were very hesitant to engage with the social problems of current sex workers because they felt it could endanger their own status as victims. Over the last few years, however, this division has visibly softened. On May 8, 2008, for instance, I was present when three former comfort women visited the camptown of Anjŏng-ri to celebrate Korean parents' day together with a group of approximately 30 aging former camptown sex workers.

5. Even after major revisions of the Status of Forces Agreement were made, to this day critics have continued to argue that South Korean law enforcement is still very much curtailed in its ability to prosecute US soldiers suspected of unlawful behavior, pointing to the fact that other countries, such as Japan and Germany, have entered into agreements that give the host country considerably more power over enforcing its own laws (Slavin 2011).

6. One direct outcome of the social movement that formed around the Yun murder was the founding of the National Campaign for Eradication of Crimes by U.S. Troops in Korea, an NGO that exists to this day, and that has played a key role in organizing the opposition to US bases in South Korea.

7. PX (which stands for Post Exchange) merchandise is sold in special stores on US military installations around the world. In South Korea, PX stores were, for a long time, a popular source for US goods, which were otherwise hard to come by.

8. The leading NGOs and activists involved in the camptowns were (and still are) endorsing a firm anti-prostitution standpoint: the casting of women involved in the sex business near US bases as victims certainly fits in to their perspective, as prostitution in their view is a business that should be banned in its entirety. A critical voice arguing against the Korean anti-prostitution perspective that is widespread among the local organizations that deal with sex work today is anthropologist Sealing Cheng, whose criticisms resemble those that can be found in the works of a number of other social scientists and activists working on prostitution worldwide (see for instance Agustin 2007; Berman 2003; Doezma 1998; Kempadoo 2005; Kempadoo and Doezma 1998; Weitzer 2000, 2005), namely, that a "focus on powerlessness and misery merely reproduces a version of the autonomous individual enshrined in civil and political rights, marginalizing discussion of economic, social, and cultural rights that importantly shape women's vulnerabilities" (Cheng 2010: 197).

9. Anthropologist Sheila Jager has explicated in detail how the nationalist left of the 1980s started to deploy a rhetoric ripe with sexualized metaphors:

The images that most frequently emerged in the context of the divided peninsula were those of the Korean woman despoiled. [...] Sexual metaphors of rape and violation were repeatedly elicited as an icon of a dislocated world used by dissident intellectuals in their portrayal of the division of their homeland. (2003: 68f)

10.

Because military brides sponsored multiple and extended family members and became the first link in long chain migrations, they have enabled the majority of Korean migration to the United States during the 1970s and early 1980s. Indeed, military brides are directly and indirectly responsible for an estimated 40 to 50 percent of all Korean immigration to the United States since 1965.

So argues Jodie Kim (2008: 292), who gives the number of 100,000 émigrés out of an estimated 1 million women employed in camptowns since US troops were permanently stationed in the country. Compare this with Cho Grace M. (2008) and Yuh Ji-Yeon (2002), who have also written on the lives of military brides in the United States.

11.

It is believed that 80 to 90 percent of the now more than 100,000 marriages between Korean women and American servicemen since 1945 have ended in divorce or separation. That's a number that is more anecdotal than empirical, but is cited by activists and community members as well as [a] small group of Korean-American academics who study Korean military brides in the United States. (Schuessler 2015)

4   Vil(l)e Encounters: Transnational Militarized Entertainment Areas on the Fringes of Korea

1. In spring 2009, I volunteered with a non-governmental organization, Turebang (My Sister's Place), for a duration of five months, usually traveling to various camptowns (located in and near Ŭijŏngbu, Tongduch'ŏn, Songt'an, Anjŏng-ri and P'aju) several times a week.

2. The drinks the soldiers buy are actually for the women. The fact that these drinks contain mostly fruit juice, to avoid quick intoxication, has also given the entertainers the questionable moniker of "juicy girl" among the soldiers.

3. Yea and Cheng's findings resonate well with what researchers have found in other geographical areas. Anthropologist Lieba Faier (2006), for instance, in her work on Filipina entertainer migrants in Japan, has also shown that notions of romantic love are the primary discursive tools deployed by migrant women in rural Japan.

4. See *Oxford English Dictionary* entry on "preoccupation" (www.oed.com).

5. This kind of phrasing was repeatedly used by people I spoke to, and also in online forums and the comments sections of blogs popular with GIs stationed

in Korea: "Love the area as it gave you a place to let off steam," one person formerly stationed near the Demilitarized Zone wrote on the (now defunct) blog rokdrop: "The girls would do any thing [sic] for you and they were not as greedy as now" (GI Korea 2009). "Why beat up on hard working soldiers or sailors who just want to have a good time [or] blow off a little steam?" another internet user asked, seeking to counter the commotion surrounding the red-light districts close to US bases on the same blog. "But hey, put all these places [that provide sex for sale] off-limits—then what will excitable young guys chained down by a curfew do? Oh yea, have sex with their fellow Soldiers, Sailors and Airmen. Let me know how well that works out" (GI Korea 2010).

6. Mazzarella, in his body of work, has made a relatively clear-cut distinction between affect and emotion that I am not fully following here. In his analysis, while feeling and emotion is a decidedly individual experience, affect (a notion that Mazzarella builds on an anthropological reading of Deleuze and Massumi) highlights embodiment and an anchoring in pre-linguistic, non-conscious, pre-social experiences that have significant effects on the way we make sense of the world.

7. See, for instance, Demick (2002), Jacoby (2002), Macintyre (2002).

8. See data provided by the South Korean National Statistics Agency at: http://www.index.go.kr/potal/main/EachDtlPageDetail.do?idx_cd=2756

9. For details, see 2001 and 2002 *Trafficking in Persons Report* at http://www.state.gov/j/tip/rls/tiprpt/.

10. Moon Seungsook reports that in the late 1980s, a handful of Filipina entertainers were working in camptown areas already (2010a: 341).

11. The current list of "off-limits" areas and establishments can be found at http://www.usfk.mil/usfk/off-limits

12. As part of the ERC-Advanced-Grant project "Overheating: The Three Crises of Globalization," I have recently conducted seven months of field research in Subic Bay, where, between September 2013 and April 2014, I explored the communal impact of a South Korean shipyard, erected on a piece of land that used to belong to the old US naval base.

13. Other non-governmental sources put the number of people living in poverty at a much higher rate. The Ibon Foundation, for instance, estimates that 56 million people live on less than 100 pesos ($2.20) a day, which amounts to more than half of the country's population (Ibon News 2014).

14. This much was confirmed to me by members of Buklod Center, a women's organization in Subic Bay that seeks to raise awareness on the issue of prostitution, and by staff of Preda Foundation, an organization assisting sexually exploited children, who all noted a significant rise in South Korean involvement in the local sex industry over recent years.

15. This facility specifically aims to be a temporary home for women rescued from the clubs nearby, with Filipinas making up the primary clientele of the shelter.

16. Raquel, in the meantime, has taken up employment as a maid in Singapore, in a job that her sister has procured for her, while Emily works part-time in a community center in a rural area of the Philippines, and is currently thinking about migrating to the Middle East for work.

17. This particular stipulation of the E6-visa states that the women's visas expire immediately if they leave the particular club they have signed a contract with. This gives rise to much abuse by local club owners, who know that their employees will try to endure bad working conditions in order to stay documented. Similar regulations also affect migrant workers in other labor sectors in Korea, with the South Korean labor authorities seeking to make a change of workplace as hard as possible. This has resulted in many human rights violations on the one hand, and in the creation of legions of undocumented workers living in precarious circumstances on the other (on this topic, see, for instance, Kim A.E. 2008; Park W. 2002).

18. Sealing Cheng notes on the widespread usage of the term "boyfriend" among Filipina entertainers:

> I put "boyfriend" in quotes because it is a term that can mean a number of things in gijichon—a casual customer, a regular customer, one of a number of customers who identify as one's "boyfriend", or a "real boyfriend" with whom one is emotionally attached. (2010: 33)

5   It'aewŏn's Suspense: Of American Dreams, Violent Nightmares, and Guilty Pleasures in the City

1. KATUSA stands for Korean Augmentation Troops to the United States Army: this program was first introduced in 1950 to strengthen the military alliance on the ground, with KATUSAs informally often functioning as "go-betweens" for US soldiers and locals (see also Moon 2010b).

2. Flower boy/flower man (kkot minam) refers to a heterosexual man of exceptional beauty. This particular type of man stands in stark contrast to the comparatively crude and tough type of man that middle-aged male Koreans who grew up during the military dictatorship represent. On the significance of the "Flower Boy," see Turnbull (2009).

3. I do not use the term district here to denote an administrative unit. The area that is commonly labelled as It'aewŏn falls under the jurisdiction of the much larger city district (gu) of Yongsan, which is further divided into neighborhoods (dong), with two of those carrying the name of It'aewŏn: Itaewon-1-dong and Itaewon-2-dong. Taken together, these two dong comprise a population of 19,316 people—which makes up less than 8 percent of the total population of 241,818 people who reside in the Yongsan district (source: June 2013 statistics—www.yongsan.go.kr).

4. *ŏnni*—older sister—is a term in Korean that expresses both familiarity and respect; it is not only used toward actual kin, but is also deployed by female speakers toward any other female speaker of older age that they are close to.

5. Lesbians frequent the area to a much lesser degree, as most lesbian meet-up places can be found in the student neighborhood of Hongdae. On the social circumstances of lesbians in Korea, see Park-Kim et al. (2007).

6. These numbers—taken from a 2010 *Korea Times* article entitled "African population in Seoul's Itaewon rises"—exclude US military personnel stationed at the Yongsan garrison from the number of foreigners living in the area. The figures should be treated with caution because a significant proportion of the African migrants in the area are undocumented and would not be included in the statistics.

7. The threat of erasure can perhaps also be seen in the way that "Hooker Hill" has disappeared from a map of the neighborhood I received at the Itaewon Tourism Information Center in It'aewŏn's subway station. The detailed map shows all the adjacent small streets in the area, but "Hooker Hill" has somehow vanished from the face of the district.

8. In *The Ritual Process*, Turner explains *communitas* in the following way:

> What is interesting about liminal phenomena for our present purposes is the blend they offer of lowliness and sacredness, of homogeneity and comradeship. We are presented, in rites of transition, with a "moment in and out of time," and in and out of secular social structure, which reveals, however fleetingly, some recognition (in symbol if not always in language) of a generalized social bond that has ceased to be and has simultaneously yet to be fragmented into a multiplicity of structural ties. (1969: 96)

9. Here, he refers to the widespread notion of the putative un-educatedness of US soldiers. For instance, Mrs. Pak, a Korean teacher at a Korean language school I attended for a while, once explained to us, her students, that "it's just that they don't have any education, that's the main problem with GIs in Korea," after I asked her what she thinks about the American soldiers in her country.

10. He is referring to the Virginia Tech massacre here, when a Korean student randomly killed 32 people at the Virginia Polytechnic Institute in April 2007, before committing suicide.

6   *Demilitarizing the Urban Entertainment Zone? Hongdae and the US Armed Forces in the Seoul Capital Area*

1. See information provided on the website of the Hongik University, at www.hongik.ac.kr.

2. "Drug" and the noisy live music it promoted became famous primarily due to the successes of one band that played there on a regular basis: Crying Nut, arguably South Korea's first and most successful punk band, formed in 1993. Their debut album, also called *Crying Nut*, sold over 100,000 copies, a phenomenal success for a band that introduced a style of music which barely had any fan base in the country prior to their appearance (Song 2010).

3. For an introduction into the phenomenon of gentrification, see Smith (1982, 1986, 2002).

4. Another crucial factor is that living in or near Hongdae has never been considered a vital necessity for those frequenting the place on a regular basis, so the rents in the neighborhood would not hinder young people with few financial resources from frequenting the neighborhood. Students, who tend to live with their parents until marriage, or who move into cheap, privately run dormitories near their universities instead, also rely heavily on Seoul's fast, inexpensive, and efficient public transportation system to get them swiftly to any place they want.

5. The "bubi bubi dance" means a form of dancing that involves a lot of body contact.

6. One very real outcome of this panic was the foundation of an organization calling themselves "Anti-English Spectrum" in 2005, a group that has made it its goal to help investigate and initiate wide discussion of the crimes of English teachers. Furthermore, in 2007, the introduction of a mandatory HIV-testing for all foreign English teachers caused much outrage among expats in the country, with those found to be HIV-positive facing immediate deportation (see Rauhala 2010; Wagner and VanVolkenburg 2012).

7. Concerning the personal toll this harassment took on them, the women involved stated: "Because of the media's selective reporting and the netizen's collective madness we are suffering incredible mental anguish and a person is receiving psychiatric treatment" (Park S. 2005, quoted in Gusts 2009).

8. In 2007, the number of foreigners in the country reached the psychologically significant benchmark of 1 million. Most of today's foreign residents are migrant workers from nearby countries or the global South.

9. The debate over the dangers of hyper-sexual(ized) Western male foreigners and their corrosive influence on the youth of Korea is noteworthy in itself, but there is one striking absence from this debate: female foreigners were broadly exempted from the discussions that followed after these scandals. To be sure, foreign women's sexuality is an issue that has raised public interest as well, but while Western foreign men are typically portrayed as sex offenders in the making, female foreigners hailing from the West, on the other hand, are put into the spotlight as ideal types of beauty and sexiness (as the earlier vignette of my encounter with Min-ho has also been hinting at). Crucial in shaping this perception has been the (over-)exposure of young foreign women on a Korean television show called *Misuda*, that aired between November 2006

and May 2010 on the KSB 2 TV channel. The format of the show involved a panel consisting of young and attractive foreign students, faced with another panel made up of an ever changing set of young Korean men who asked the foreign women questions that usually revolved around their (love) lives in Korea. While the panel featured foreign women from all kinds of ethnic backgrounds and countries of origin, the "stars" of the show were clearly the white or half-white women, some of whom ended up with well-paid contracts for commercials, record deals, or were even able to launch independent acting careers.

An Uzbek woman called Jamilya, for instance, achieved further TV fame when, after a brief stint on *Misuda*, she was cast as the lead in a show called *Sexy Mong Returns*. Hunting down sex offenders while wearing short skirts and high heels, Jamilya's trademark on the show was the lollipop that she was continuously sucking on while going after the bad guys of each episode. The very first episode of this thinly veiled soft porn (first aired in the spring of 2008) featured Jamilya and her Korean friends bringing to justice a group of English teachers who were portrayed as regularly taking advantage of drunk Korean women in the club scene of Hongdae. "Korean girls are good," we hear one of the teachers say, "they are so easy." And another one admits, "I just know about Kimchi and Korean girls." At the end of the episode, we see Jamilya and her friends bent over a newspaper with the headline of "English teachers caught up in sexy party scandal," a clear reference to the actual controversy surrounding the "sexy party" pictures taken at a Hongdae club a few years earlier. In this way, the image of the Western woman was made complicit in the symbolic purge of the Western male that was understood as polluting the entertainment districts of Seoul, while Jamilya was simultaneously made available as a sex object for a Korean male audience.

10. Up to $10,000 per year is the standard cost of tuition fees at many of Seoul's more prestigious universities nowadays, and scholarships are very hard to come by in the extremely competitive climate of higher education in South Korea.

11. This matter was also picked up in an article on Hongdae's punk scene by a journalist writing for *Vice Magazine*. The singer of one band, for instance, is quoted in the article as saying "Korean society is just a bunch of fucking nationalists. We are all brought up to hate. Eighty percent of Korean men are stupid fascists." A girl called Rosa, on the other hand, argues in the same piece that punk is her escape from and "retaliation against the slave-like work ethic, the lack of individuality, the lack of any real culture. All people do is drink and fight." And another punk, Ki, relayed that his own:

> [m]ilitary service was fucking awful. [...] They teach you how to kill a man. You can't get out of it—if you try to avoid it you lose your Korean citizenship altogether. We're a nation of potential murderers. When I was

in the service there was a big riot at the US embassy and protesters were burning American flags. I was sent in there to repress them. I may not have wanted to, but if I didn't do my mandatory duty to my nation I'd be thrown into jail, so what was I supposed to do? (Hoban 2009)

12. Hyŏn-jun, in his letter to military officials in which he would eventually declare himself a conscientious objector, made his anarchist convictions the central point that had driven him toward refusing to serve the military: "A human being," he writes, "turns into its true self from the moment it *wants* to become something. I wanted to become an anarchist ever since I was young. This is not because I was impressed by a new kind of Western knowledge and wanted to follow it, but it was the outcome of looking for my own goodness, during which I found a lot of similar ideas amongst those anarchists that have existed before me. If someone were to ask me what my ideals are—to be able to answer that without complexity I am borrowing the mask of anarchism."

13. *Temporary Autonomous Zones* (1991) is the title of a book by anarchist writer Peter Lamborn Wilson, who authored it under the pseudonym of Hakim Bey. In it, he explores the possibilities for the creation of interstitial spaces overlooked by repressive state actors; spaces in which alternative political, social and sexual relationships may be temporarily shaped that allow the people involved in them to evade forms of control from above.

References

Abelmann, N. (1996) *Echoes of the Past, Epics of Dissent: A South Korean Social Movement* (Berkeley: University of California Press).

—— (1997) "Reorganizing and recapturing dissent in 1990s South Korea," in Fox, R. and Starn, O. (eds.) *Between Resistance and Revolution: Cultural Politics and Social Protest* (New Brunswick, NJ: Rutgers University Press), pp. 251–281.

ABS-CBN News (2015) "Poverty incidence rises in Philippines," ABS-CBN News, June 3. Available at: http://www.abs-cbnnews.com/business/03/06/15/poverty-incidence-rises-philippines (accessed July 9, 2015).

Adesnik, D.A. and Kim, S. (2008) *If at First You Don't Succeed: The Puzzle of South Korea's Democratic Transition.* CDDRL Working Papers 83 (Stanford: CDDRL).

Agustin, L. (2007) *Sex at the Margins: Migration, Labor Markets, and the Rescue Industry* (London: Zed Books).

Anderson, B. (1991 [1983]) *Imagined Communities* (London: Verso).

Angst, L. (1995) "The rape of a school girl: discourses of power and women's lives in Okinawa," in Hein, L.E. and Selden, M. (eds.) *Islands of Discontent: Okinawan Responses to Japanese and American Power* (Lanham, MD: Rowman & Littlefield), pp. 135–157.

—— (2001) "The sacrifice of a schoolgirl: the 1995 rape case, discourses of power, and women's lives in Okinawa," *Critical Asian Studies* 33(2): 243–266.

Appadurai, A. (1996) *Modernity at Large: Cultural Dimensions of Globalization.* (Minneapolis: University of Minnesota Press).

—— (2013) *The Future as Cultural Fact: Essays on the Global Condition* (London: Verso).

Atkins, E.T. (2010) *Primitive Selves: Koreana in the Japanese Colonial Gaze, 1910–1945* (Berkeley: University of California Press).

Baca, G. (2010) *Conjuring Crisis: Racism and Civil Rights in a Southern Military City* (New Brunswick, NJ: Rutgers University Press).

BBC (2007) "Eviction village: a farmer's tale," February 27. Available at: http://news.bbc.co.uk/2/hi/asia-pacific/6389553.stm (accessed January 9, 2016).

Benford, R.D. and Snow, D.A. (2000) "Framing processes and social movements: an overview and assessment," *Annual Review of Sociology* 26: 611–639.

Berman, J. (2003) "(Un)popular strangers and crises (un)bounded: discourses of sex-trafficking, the European political community and the panicked state of the modern state," *European Journal of International Relations* 9(1): 37–86.

Bermudez, J. (2001) *The Armed Forces of North Korea* (New York: I.B. Tauris).

Bey, H. (1991) *A.Z.: The Temporary Autonomous Zone, Ontological Anarchy, Poetic Terrorism* (New York: Autonomedia).

Brazinsky, G. (2007) *Nation Building in South Korea: Koreans, Americans, and the Making of a Democracy* (Chapel Hill, NC: University of North Carolina Press).

Brenner, N. (1999) "Globalisation as reterritorialisation: the re-scaling of urban governance in the European Union," *Urban Studies* 36(3): 431–451.

Brighenti, A.M. (2010) "On territorology: towards a general science of territory," *Theory, Culture & Society* 27(1): 52–72.

Burawoy, M., Blum, J.A., George S., Gille, Z. et al. (eds.) (2000) *Global Ethnography: Forces, Connections and Imaginations in a Postmodern World* (Berkeley: University of California Press).

Capaccio, A. and Gaouette, N. (2014) "U.S. adding 800 troops for South Korea citing rebalance," *Bloomberg*, January 7. Available at: http://www.bloomberg.com/news/articles/2014-01-07/u-s-adding-800-troops-for-south-korea-citing-rebalance (accessed July 9, 2015).

Caprio, M. (2009) *Japanese Assimilation Policies in Colonial Korea, 1910–1945* (Seattle: University of Washington Press).

Caton, S.C. (1999) "Anger be now thy song: the anthropology of an event," Occasional Papers of the School of Social Science No. 5. Available at: https://www.sss.ias.edu/files/papers/paperfive.pdf (accessed July 9, 2015).

Ceuster, K.D. (2002) "The nation exorcised: the historiography of collaboration in South Korea," *Korean Studies* 25(2): 207–242.

Chai, A.Y. (1993) "Asian-Pacific feminist coalition politics: the Chongshindae / Jugu,nianfu ('Comfort Women') Movement," *Korean Studies* 17: 67–91.

Chang, D. (2006) "Samsung moves: a portrait of struggles," in Chang, D. (ed.) *Labor in Globalising Asian Corporations* (Hong Kong: Asia Monitor Resource Center).

Chang, P.Y. (2008) "Unintended consequences of repression: alliance formation in South Korea's democracy movement (1970–1979)," *Social Forces* 87(2): 651–677.

Chen, J. (1994) *China's Road to the Korean War: The Making of the Sino-American Confrontation* (New York: Columbia University Press).

Cheng, S. (2010) *On the Move for Love: Migrant Entertainers and the U.S. Military in South Korea* (Philadelphia: University of Pennsylvania Press).

—— (2015) "Beyond trafficking and slavery: Filipina entertainers and South Korean anti-trafficking laws," *Open Democracy*, 31 March. Available at: https://www.opendemocracy.net/beyondslavery/sealing-cheng/filipina-entertainers-and-south-korean-antitrafficking-laws (accessed July 9, 2015).

Cho, G.M. (2008) *Haunting the Korean Diaspora: Shame, Secrecy, and the Forgotten War* (Minneapolis: University of Minnesota Press).

Cho, M. (2007) *Construction of Hong-dae Cultural District: Cultural Place, Cultural Policy and Cultural Politics* (PhD dissertation, University of Bielefeld).

Cho, S. (2013) "United States Forces Korea's (USFK) crisis communication strategies and crisis responses: the case of two Korean school girls' death," *International Journal of Contents* 9(1): 98–103.

Cho, Y. (2003) *A Single Spark: The Biography of Chun Tae-il* (Seoul: Dolbegae Publishers).

Choe, S. (2009) "Ex-prostitutes say South Korea enabled sex trade near U.S. military bases," *The New York Times*, January 8. Available at: http://www.nytimes. com/2009/01/08/world/asia/08iht-08korea.19174342.html (accessed July 9, 2015).

Choi, C. (2005) "Kŭdŭrŭn urirŭl ch'angnyŏ, yanggongju, p'ochuro mollakatta," *OhmyNews*, January 20. Available at: http://news.naver.com/main/read. nhn?mode=LSD&mid=sec&sid1=102&oid=047&aid=0000057135 (accessed July 9, 2015).

Choi, L. (2012) *The Foreign Policy of Park Chunghee: 1968–1979* (PhD thesis, London School of Economics). Available at: http://etheses.lse.ac.uk/506/1/Choi_The%20 Foreign%20Policy%20of%20Park%20Chunghee.pdf (accessed July 11, 2015).

Chosun Ilbo (2005a) "Indie flashers planned exposure in advance," *Chosun Ilbo*, April 8. Available at: http://english.chosun.com/site/data/html_ dir/2005/08/04/2005080461027.html (accessed July 9, 2015).

Chosun Ilbo (2005b) "Punk rockers' privates in affront to Korea's 'bourgeois,'" *Chosun Ilbo*, July 31. Available at http://english.chosun.com/site/data/html_ dir/2005/07/31/2005073161005.html (accessed July 9, 2015).

Chosun Ilbo (2005c) "Seoul mayor blasted for authoritarian mindset," *Chosun Ilbo*, August 2. Available at: http://english.chosun.com/site/data/html_ dir/2005/08/02/2005080261010.html (accessed July 9, 2015).

Chun, S. (2002) "Sorry, soldier, can't let you in," *JoongAng Daily*, December 6. Available at: http://koreajoongangdaily.joins.com/news/article/article. aspx?aid=1912029 (accessed July (, 2015).

—— (2004) "A taste of Russia in heart of Seoul," *JoongAng Daily*, August 10. Available at: http://koreajoongangdaily.joins.com/news/article/Article.aspx?aid=2453011 (accessed July 9, 2015).

Chung, H. (n.d.) *A Shot in the Dark: Korean Military Brides in America* (online book). Available at: http://issuu.com/hsmchung/docs/kmb/1 (accessed July 9, 2015).

Chung, J.H. (2006) *Between Ally and Partner: Korea–China Relations and the United States* (New York: Columbia University Press).

Cocking, J.M. (1991) *Imagination: A Study in the History of Ideas* (London: Routledge).

Cohn, C. (2012) *Women and Wars: Contested Histories, Uncertain Futures.* (Cambridge: Polity).

Congressional Budget Office (1997) *The Role of Foreign Aid in Development: South Korea and the Philippines.* Available at: https://www.cbo.gov/publication/14518 (accessed July 9, 2015).

Cooley, A. (2005) "Democratization and the contested politics of U.S. military bases in Korea," *IRI Review* 10(2): 201–232.

—— (2008) "US bases and democratization in Central Asia," *Orbis* 52(1): 65–90.

Cooper, F. and Stoler, L. (eds.) (1997) *Tensions of Empire: Colonial Cultures in a Bourgeois World* (Berkeley: University of California Press).

Crapanzano, V. (2004) *Imaginative Horizons: An Essay in Literary Philosophical Anthropology* (Chicago: University of Chicago Press).

Cumings, B. (1981, 1990) *The Origins of the Korean War* (2 vols) (Princeton, NJ: Princeton University Press).

—— (1997) *Korea's Place in the Sun: A Modern History* (New York: W.W. Norton).

—— (2003) "Colonial formations and deformations: Korea, Taiwan and Vietnam," in Duara, P. (ed.) *Decolonization: Perspectives from Now and Then* (New York: Routledge), pp. 278–297.

—— (2004) *North Korea: Another Country* (New York: The New Press).

—— (2005) "The structural basis of 'anti-Americanism' in the Republic of Korea," in Steinberg, D.I. and Gallucci, R. (eds.) *Korean Attitudes Toward the United States: Changing Dynamics* (Armonk, NY: M.E. Sharpe Inc.), pp. 91–114.

Das, V. (1997) *Critical Events: An Anthropological Perspective on Contemporary India* (Oxford: Oxford University).

Davis, M. (2006) *Planet of Slums* (London: Verso).

Day, S. (2007) *On the Game: Women and Sex Work* (London: Pluto Press).

Day, S. and Ward, H. (2004) *Sex Work, Mobility and Health* (New York: Routledge).

Demick, B. (2002) "Off-base behavior in Korea," *Los Angeles Times*, September 26. Available at: http://articles.latimes.com/2002/sep/26/world/fg-barwomen26 (accessed July 9, 2015).

Doak, K.M. (2008) "Narrating China, ordering East Asia: the discourse on nation and ethnicity in Imperial Japan," *Journal of the Washington Institute of China Studies* 3(1): 1–24.

Doezma, J. (1998) "Forced to choose: beyond the voluntary v. forced prostitution dichotomy," in Kempadoo, K. and Doezema, J. (eds.) *Global Sex Workers: Rights, Resistance, and Redefinition* (New York: Routledge), pp. 34–50.

Donovan, B. and Harcourt, C. (2005) "The many faces of sex work," *Sexual Health* 81(3): 11–28.

Drennan, W.H. (2005) "The tipping point: Kwangju, Mai, 1980," in Steinberg, D.I. and Gallucci, R. (eds.) *Korean Attitudes Toward the United States: Changing Dynamics* (Armonk, NY: M.E. Sharpe Inc.), pp. 280–306.

Eckert, C.J. (2000) *Offspring of Empire: The Koch'ang Kims and the Colonial Origins of Korean Capitalism 1876–1945* (Seattle: University of Washington Press).

Elshtain, J.B. (1987) *Women and War* (New York: Basic Books).

Em, H. (1999) "Nationalism, post-nationalism, and Shin Ch'ae-ho," *Korea Journal* 39(2): 283–317.

Enloe, C. (1983) *Does Khaki Become You? The Militarization of Women's Lives* (Berkeley: University of California Press).

—— (1989) *Bananas, Beaches and Bases: Making Feminist Sense of International Politics* (London: Pandora).

—— (1992) "It takes two," in Sturdevant, S.P. and Stoltzfus, B. (eds.) *Let the Good Times Roll: Prostitution and the U.S. Military in Asia* (New York: The New Press), pp. 22–27.

—— (2000) *Maneuvers: The International Politics of Militarizing Women's Lives* (Berkeley: University of California Press).

Faier, L. (2006) "Filipina migrants in rural Japan and their professions of love," *American Ethnologist* 34(1): 148–162.

Feinerman, J.V. (2005) "The U.S.–Korean Status of Forces Agreement as a source of continuing Korean anti-American attitudes," in Steinberg, D.I. and Gallucci, R. (eds.) *Korean Attitudes Toward the United States: Changing Dynamics* (Armonk, NY: M.E. Sharpe Inc.), pp. 196–218.

Flack, T.D. (2007) "Hongdae district is placed off-limits to SOFA personnel at night," *Stars and Stripes*, February 3. Available at: http://www.stripes.com/news/hongdae-district-is-placed-off-limits-to-sofa-personnel-at-night-1.59824 (accessed July 9, 2015).

Forte, M. (2011) *The New Imperialism 2: Interventionism, Information Warfare, and the Military-Academic Complex* (Montreal: Alert Press).

Frese, P.R. and Harrell, M.C. (eds.) (2003) *Anthropology and the United States Military: Coming of Age in the Twenty-first Century* (New York: Palgrave Macmillan).

Friedmann, J. and Miller, J. (1965) "The urban field," *Journal of the American Institute of Planners* 31(4): 312–320.

Gateward, F. (2007) "Waiting to exhale: the colonial experience and the trouble with my own breathing," in Gateward, F. (ed.) *Seoul Searching: Culture and Identity in Contemporary Korean Cinema* (Albany: State University of New York Press), pp. 191–217.

Gellner, E. (1983) *Nations and Nationalism* (Ithaca, NY: Cornell University Press).

GI Korea (2009) "Ville memories: Changpa-ri, Korea – then and now," *rokdrop* (online blog), April 2. Available at: http://rokdrop.com/2009/04/02/ville-memories-changpa-ri-then-now/ (accessed September 9, 2011, now defunct).

GI Korea (2010) "US Congress gets involved in South Korea juicy bar issue," *rokdrop* (online blog) April 23. Available at: http://rokdrop.com/2010/04/23/us-congress-gets-involved-in-south-korea-juicy-bar-issue/ (accessed September 9, 2011, now defunct).

Gibson, L. (2014) "Guest editorial: anthropology and imagination," *Sites: A Journal of Social Anthropology and Cultural Studies* 11(1): 3–14.

Gill, L. (2007) "Anthropology goes to war, again," *Focaal* 50: 139–145.

Gille, Z. and Ó Riain, S. (2002) "Global ethnography," *Annual Review of Sociology* 28: 271–295.

Glassman, J. and Choi, Y. (2014) "The chaebol and the US military-industrial complex: Cold War geopolitical economy and South Korean industrialization," *Environment and Planning* 46: 1160–1180.

Goffman, E. (1974) *Frame Analysis: An Essay on the Organization of Experience* (London: Harper & Row).

—— (1990 [1963]) *Stigma: Notes on the Management of Spoiled Identity* (London: Penguin Books).

González, R.J. (2009) *American Counterinsurgency: Human Science and the Human Terrain* (Chicago: Prickly Paradigm Press).

Graeber, D. (2012) "Dead zones of the Imagination: on violence, bureaucracy and interpretive labour," *HAU: Journal of Ethnographic Theory* 2(2): 105–128.

—— (2015) *The Utopia of Rules: On Technology, Stupidity, and the Secret Joys of Bureaucracy* (New York: Melville House).

Grassiani, E. (2013) *Soldiering Under Occupation: Moral Numbing among Israeli Conscripts during the Al-Aqsa Intifada* (Oxford: Berghahn Books).

Gregg, D. (1999) "Park Chung Hee," *Time*, August 23. Available at: http://content. time.com/time/world/article/0,8599,2054405,00.html (accessed July 9, 2015).

Gupta, A. and Ferguson, J. (eds.) (1997) *Culture, Power, Place: Explorations in Critical Anthropology* (London: Duke University Press).

Gusterson, H. (2007) "Anthropology and militarism," *Annual Review of Anthropology* 36: 155–175.

Gusts (2005) "Pants dropping leads to trail of media droppings," *Gusts of Popular Feeling* (online blog) August 22. Available at: http://populargusts.blogspot. co.at/2005/08/pants-dropping-leads-to-trail-of-media.html (accessed July 9, 2015).

Gusts (2009) "A new classic," *Gusts of Popular Feeling* (online blog) July 1. Available at: http://populargusts.blogspot.co.at/2009/07/new-classic.html (accessed July 9, 2015).

Gutmann, M. and Lutz, C. (2010) *Breaking Ranks: Iraq Veterans Speak Out Against the War* (Berkeley: University of California Press).

Han, G. (2003) "African migrant workers' views of Korean people and culture," *Korea Journal* 43(1): 154–173.

Han, J. (2001) *Yanggongju – "die zeitweiligen Honeys" der US-amerikanischen Soldaten in Südkorea* (Münster: LIT Verlag).

Han, S.J. (2005) "Imitating the colonizers: the legacy of the disciplining state from Manchukuo to South Korea," *Asia Pacific Journal: Japan Focus*, July 10. Available at: http://www.japanfocus.org/-Suk_Jung-Han/1885 (accessed July 9, 2015).

Han, S.K. (2008) "Breadth and depth of unity among chaebol families in Korea," *Korean Journal of Sociology* 42(4): 1–25.

Hankyoreh (2009) "Seoul ranks highest in population density among OECD countries," *Hankyoreh* , December 15. Available at: http://www.hani.co.kr/arti/ english_edition/e_international/393438.html (accessed July 9, 2015).

Harvey, D. (2008) "The right to the city," *New Left Review* 53: 23–40.

—— (2012) *Rebel Cities: From the Right to the City to Urban Revolutions* (London: Verso).

Havely, J. (2003) "Korea's DMZ: 'scariest place on earth,'" *CNN Hong Kong*, August 28. Available at: http://edition.cnn.com/2003/WORLD/asiapcf/east/04/22/ koreas.dmz/ (accessed July 9, 2015).

Hein, L. (1999) "Savage irony: the imaginative power of the 'military comfort women' in the 1990s," *Gender and History* 11(2): 336–372.

Hewamanne, S. (2013) "The war zone in my heart: the occupation of southern Sri Lanka," in Visweswaran, K. (ed.) *Everyday Occupations: Experiencing Militarism in*

South Asia and the Middle East (Philadelphia: University of Pennsylvania Press), pp. 60–84.

Hicks, G. (1995) *The Comfort Women: Japan's Brutal Regime of Enforced Prostitution in the Second World War* (New York: W.W. Norton).

Hoban, A. (2009) "South Korea – punk here is like it is everywhere else," *Vice Magazine*, July 5. Available at: http://vice.typepad.com/vice_magazine/2009/05/south-korea-total-angst.html#more%3Cbr/%3E (accessed July 12, 2015).

Hobsbawm, E. (1991) *Nations and Nationalism since 1780: Programme, Myth, Reality* (Cambridge: Cambridge University Press).

Hobsbawm, E. and Ranger, T. (1983) *The Invention of Tradition* (Cambridge: Cambridge University Press).

Hoehn, M. and Moon, S. (eds.) (2010) *Over There: Living with the U.S. Military Empire from World War Two to the Present* (Durham, NC: Duke University Press).

Hugh, T. (2005) "Development as devolution: Nam Chŏng-hyŏn and the 'Land of Excrement' incident," *Journal of Korean Studies* 10(1): 29–57.

Ibon News (2014) "Economy under the Aquino administration: worsening exclusivity," *Ibon News*, 28 July. Available at: http://ibon.org/ibon_articles.php?id=424 (accessed July 9, 2015).

Isozaki, N. (2002) "South Korea: advocacy for democratization," in Shigetomi, S. (ed.) *The State and NGOs: Perspectives from Asia* (Singapore: Institute of Southeast Asian Studies), pp. 288–310.

Jacoby, M. (2002) "Does U.S. abet Korean sex trade?" *St. Petersburg Times*, December 9. Available at: http://www.sptimes.com/2002/12/09/Worldandnation/Does_US_abet_Korean_s.shtml (accessed July 9, 2015).

Jager, S. (2003) *Narratives of Nation-Building in Korea: A Genealogy of Patriotism* (Armonk, NY: M.E. Sharpe).

Janelli, R.L. (1993) *Making Capitalism: The Social and Cultural Construction of a South Korean Conglomerate* (Stanford, CA: Stanford University Press).

Jaschik, S. (2015) "Embedded conflicts," *Inside Higher Ed*, July 7. Available at: https://www.insidehighered.com/news/2015/07/07/army-shuts-down-controversial-human-terrain-system-criticized-many-anthropologists (accessed July 11, 2015).

Jeon, B. (2005) *Queer Mapping in Seoul: Seoul Until Now!* (exhibition catalogue) (Copenhagen: Charlottenborg Udstillingsbygning), pp. 70–77.

Jin, H. (2005) "Police to probe punk band: was TV flashing planned?" *Korea Times*, August 2. Available at: http://www.asiamedia.ucla.edu/print.asp?parentid=27557 (accessed July 9, 2015).

Jin, K. (2014) "China's charm offensive toward South Korea," *The Diplomat*, July 8. Available at: http://thediplomat.com/2014/07/chinas-charm-offensive-toward-south-korea/ (accessed July 9, 2015).

Johnson, C. (2004) *The Sorrows of Empire: Militarism, Secrecy, and the End of the Republic* (New York: Metropolitan Books).

JoonAng Daily (2005) "Rude rockers get suspended sentences for raunchy stunt," *JoongAng Daily*, September 27. Available at: http://joongangdaily.joins.com/article/view.asp?aid=2623071 (accessed July 9, 2015).

Kalinowski, T. and Cho, H. (2012) "Korea's search for a global role between hard economic interests and soft power," *European Journal of Development Research* 24: 242–260.

Kang, M. (1996) *The Korean Business Conglomerate: Chaebol Then and Now*, Center for Korean Studies Korea Research Monograph no. 21 (Berkeley: Institute of East Asian Studies, University of California).

Katsiaficas, G. (2006) "Neoliberalism and the Gwangju uprising," *Korea Policy Review* II. Available at: http://www.eroseffect.com/articles/neoliberalismgwangju.htm#_ednref71 (accessed July 9, 2015).

—— (2009) "South Korea's rollback of democracy," 25 May. Available at: http://www.eroseffect.com/articles/rollback.htm (accessed July 9, 2015).

Katsiaficas, G. and Na, K.C. (2006) *South Korean Democracy: Legacy of the Kwangju Uprising* (New York: Routledge).

Kempadoo, K. (2005) *Trafficking and Prostitution Reconsidered: New Perspectives on Migration, Sex Work, and Human Rights* (London: Paradigm Publishers).

Kempadoo, K. and Doezma, J. (1998) *Global Sex Workers: Rights, Resistance, and Redefinition.* (New York: Routledge).

Kern, T. (2005) "Anti-Americanism in South Korea: from structural cleavages to protest," *Korea Journal* 45(1): 257–288.

Kim, A.E. (2008) "Global migration and South Korea: foreign workers, foreign brides and the making of a multicultural society," *Ethnic and Racial Issues* 31(1): 70–92.

Kim, B. (2007) "Dongducheon, and Korea, now where to? Yesterday, today, and tomorrow of the U.S. Army base town," in *Dongducheon – A Walk to Remember, A Walk to Envision* (exhibition leaflet) (Seoul: Insa Art Space), pp. 20–27.

Kim, B.K. and Im, H.B. (2001) "Crony capitalism in South Korea, Thailand and Taiwan: myth and reality," *Journal of East Asian Studies* 1(1): 5–52.

Kim, D.C. (2006) "Growth and crisis of the Korean citizens' movement," *Korea Journal* 46(2): 99–128.

Kim, E. (2004) "Itaewon as an alien space within the nation-state and a place in the globalization era," *Korea Journal* 44(3): 34–64.

Kim, E. and Park, G. (2011) "The Chaebol," in Kim, B. and Vogel, E.F. (eds.) *The Park Chung Hee Era: The Transformation of South Korea* (Cambridge, MA: Harvard University Press), pp. 265–294.

Kim, H.A. (2005) *Korea's Development under Park Chung Hee* (London: RoutledgeCurzon).

Kim, H.R. (2003) "Unraveling civil society in South Korea: old discourses and new visions," in Schak, D.C. and Hudson, W. (eds.) *Civil Society in Asia* (Burlington, VT: Ashgate), pp. 192–209.

Kim, H.R. and McNeal, D.K. (2007) "From state-centric to negotiated governance: NGOs as policy entrepreneurs in South Korea," in Weller, R.P. (ed.) *Civil Life, Globalization, and Political Change in Asia* (New York: Routledge), pp. 95–109.

Kim, H.S. (2009 [1997]) *The Women Outside: Korean Women and the U.S. Military* (New York: Third World Newsreel). Available at: http://www.twn.org/catalog/guides/WomenOutside_StudyGuide.pdf (accessed July 9, 2015).

—— (1998) "Yanggongju as an allegory of the nation: images of working-class women in popular and radical texts," in Kim, E.H. and Choi, C. (eds.) *Dangerous Women: Gender and Korean Nationalism* (London: Routledge), pp.175–202.

Kim, I.K. (2010) "Socioeconomic concentration in the Seoul Metropolitan Area and its implications in the urbanization process of Korea," *Korean Journal of Sociology* 44(3): 111–123.

Kim, J. (2007) "Queer cultural movements and local counterpublics of sexuality: a case of Seoul Queer Films and Videos Festival," *Inter-Asia Cultural Studies* 8(4): 617–633.

Kim, J., Moon, Y., and Kang, B. (2006) "The farewell to mediation for the national security in South Korea: The guns of May in Pyeongtaek City" (conference paper, presented at Asia-Pacific Mediation Forum, Suva, Fiji). Available at: http://www.apmec.unisa.edu.au/apmf/2006/papers/kim-moon-kang.pdf (accessed July 9, 2015).

Kim, J. (2001) "From "American gentlemen" to "Americans": changing perceptions of the United States in South Korea in recent years," *Korea Journal* 41(4): 172–198.

Kim, J. (2008) "'I'm not here, if this doesn't happen': the Korean War and Cold War epistemologies in Susan Choi's *The Foreign Student* and Heinz Insu Fenkl's *Memories of My Ghost Brother*," *Journal of Asian American Studies* 11(3): 279–302.

Kim, K. and Gil Y. (2013) "Huge increase in US Troops in South Korea," *Hankyoreh*, March 21. Available at: http://www.hani.co.kr/arti/english_edition/e_international/579051.html (accessed July 9, 2015).

Kim, M. and Yang, J. (2010) "The 'East Berlin spy incident' and the diplomatic relations between South Korea and West Germany, 1967–1970: why did the diplomatic strains last so long?," in Cuc, C. (ed.) *Multidisciplinary Perspectives in Korean Studies* (Proceedings of the 7th Korean Studies Graduate Students Convention in Europe), (Cluj-Napoca: Asian Studies Department Babes-Bolyai University).

Kim, N. (2008) *Imperial Citizens: Koreans and Race from Seoul to L.A.* (Stanford, CA: Stanford University Press).

Kim, P. and Shin, H. (2010) "The birth of "rok": cultural imperialism, nationalism, and the glocalization of rock music in South Korea, 1964–1975," *Positions. East Asia Cultures Critique* 18(1): 199–230.

Kim, T. (2005) "Music show canceled after indecent exposure," *The Korea Times*, July 31. Available at: http://web.international.ucla.edu/asia/article/27461 (accessed July 9, 2015).

Kim, Y. and Hahn, S. (2006) "Homosexuality in ancient and modern Korea," *Culture, Health & Sexuality* 8(1): 59–65.

Kirk, D. (2013) *Okinawa and Jeju: Bases of Discontent* (New York: Palgrave Macmillan).

Koehler, R. (2005a) "Korean netizens blast foreign English teacher site," *The Marmot's Hole* (online blog) January 12. Available at: http://www.rjkoehler. com/2005/01/12/korean-netizens-blast-foreign-english-teacher-site/ (accessed July 9, 2015).

Koehler, R. (2005b) "English Spectrum gate continues!" *The Marmot's Hole* (online blog) January 14. Available at: http://www.rjkoehler.com/2005/01/14/english-spectrum-gate-continues/ (accessed July 9, 2015).

Koehler, R. (2005c) "Hongik U – a hookup paradise for foreign men and Korean women," *The Marmot's Hole* (online blog) August 9. Available at: http://www. rjkoehler.com/2005/08/09/hongik-u-a-hookup-paradise-for-foreign-men-and-korean-women/ (accessed July 15, 2015).

Koo, H. (2001) *Korean Workers: The Culture and Politics of Class Formation* (Ithaca, NY: Cornell University Press).

Korea Times (2007) "10 surprises of Korea (1950–2007)," *Korea Times*, November 23. Available at: http://www.koreatimes.co.kr/www/news/include/print. asp?newsIdx=12432 (accessed July 9, 2015).

Korea Times (2010) "African population in Seoul's Itaewon rises," *Korea Times*, April 13. Available at: http://www.koreatimes.co.kr/www/news/nation/2010/04/ 113_64096.html (accessed July 9, 2015).

Korea Tourism Association (n.d.) "Itaewon." Available at: http://english.visitkorea. or.kr/enu/SH/SH_EN_7_2_6_1.jsp (accessed July 9, 2015).

Krueger, A.O. and Yoo, J. (2002) "Chaebol capitalism and the currency-financial crisis in Korea," in Edwards, S. and Frankel, J.A. (eds.) *Preventing Currency Crises in Emerging Markets* (Chicago: University of Chicago Press), pp. 601–661.

Kwon, S. and O'Donnell, M. (2001) *The Chaebol and Labour in Korea: The Development of Management Strategy in Hyundai* (London: Routledge).

Lacsamana, A.E. (2011) "Empire on trial: the Subic rape case and the struggle for Philippine women's liberation," *Works and Days* 29(57/58): 203–215.

Lankov, A. (2003) *From Stalin to Kim Il Sung: The Formation of North Korea, 1945–1960* (London: Rutgers University Press).

—— (2007) *The Dawn of Modern Korea* (Seoul: EunHaeng Namu).

—— (2011) "Tragic end of Communist-turned-politician Cho Bong-am," *Korea Times*, January 9. Available at: http://www.koreatimes.co.kr/www/news/ nation/2011/01/113_79367.html (accessed July 9, 2015).

Lee, C. (2015) "Korean dream shattered by lies, sex trade coercion," *Korea Herald*, June 19. Available at: http://www.koreaherald.com/view.php?ud=20150615000845 (accessed July 9, 2015).

Lee, H.K. (1996) "NGOs in Korea," in Yamamoto, T. (ed.) *Emerging Civil Society in the Asia Pacific Community* (Seattle: Unversity of Washington Press), pp. 161–165.

Lee, J. (2009) "Surrogate military, subimperialism, and masculinity: South Korea in the Vietnam War, 1965–73," *positions* 17(3): 655–682.

—— (2010) *Service Economies: Militarism, Sex Work, and Migrant Labor in South Korea* (Minneapolis: University of Minnesota Press).

Lee, J. (2012) "Micro-dynamics of protests: the political and cultural conditions for anti-U.S. beef protests in South Korea," *Sociological Perspectives* 55(3): 399–420.

Lee, M. (2008) "Mixed race peoples in the Korean national imaginary and family," *Korean Studies* 32: 56–85.

Lee, M. (2005) "Openly revealing a secret life," *Korea JoongAng Daily*, July 31. Available at: http://koreajoongangdaily.joins.com/news/article/article.aspx?aid=2600608 (accessed July 11, 2015).

Lee, M. (2004) "The landscape of club culture and identity politics: focusing on the club culture in the Hongdae area of Seoul," *Korea Journal* 44(3): 65–107.

Lee, N. (2002) "Anticommunism, North Korea, and human rights in South Korea: 'Orientalist' discourse and construction of South Korean identity," in Bradley, M.P. and Petro, P. (eds.) *Truth Claims: Representation and Human Rights* (London: Rutgers University Press), pp. 43–72.

—— (2007) *The Making of Minjung: Democracy and Politics of Representation in South Korea* (Ithaca, NY: Cornell University Press).

Lee, N.Y. (2007) "The construction of military prostitution in South Korea during the US military rule, 1945–1948," *Feminist Studies* 33(3): 453–481.

Lee, S., Kim, S. and Wainwright, J. (2010) "Mad cow militancy: neoliberal hegemony and social resistance in South Korea," *Political Geography* 29(7): 359–369.

Lie, J. (1998) *Han Unbound: The Political Economy of South Korea* (Stanford, CA: Stanford University Press).

Link, B.G. and Phelan, J.C. (2001) "Conceptualizing stigma," *Annual Review of Sociology* 27: 363–385.

Lutz, C. (2001) *Homefront: A Military City and the American 20th Century* (Boston, MA: Beacon Press).

—— (2002a) "Making war at home in the United States: militarization and the current crisis," *American Anthropologist* 104(3): 723–735.

—— (2002b) "The wars less known," *South Atlantic Quarterly* 101(2): 285–296.

—— (2006) "Empire is in the details," *American Ethnologist* 33(4): 593–611.

—— (ed.) (2009a) *The Bases of Empire: The Global Struggle Against U.S. Military Posts* (London: Pluto Press).

—— (2009b) "US foreign military bases: the edge and essence of empire," in Susser, I. and Maskovsky, J. (eds.) *Rethinking America* (Boulder, CO: Paradigm Publishers), pp. 15–30.

—— (2010) "US military bases on Guam in global perspective," *Asia-Pacific Journal: Japan Focus*, July 26. Available at: http://www.japanfocus.org/-catherine-lutz/3389/article.html (accessed July 11, 2015).

Macintyre, D. (2002) "Base instincts," *Time Magazine*, August 5. Available at: http://content.time.com/time/magazine/article/0,9171,333899,00.html (accessed July 9, 2015).

Majic, S. (2014) *Sex Work Politics: From Protest to Service Provision* (Philadelphia, PA: University of Pennsylvania Press).

Mann, M. (2003) *Incoherent Empire* (London: Verso Books).

Mason, C. (2009) "Status of Forces Agreement (SOFA): what is it, and how has it been utilized?" (Congressional Research Service Report for Congress). Available at: http://www.fas.org/sgp/crs/natsec/RL34531.pdf (accessed July 9, 2015).

Mazzarella, W. (2009) "Affect: what is it good for?" in Dube, S. (ed.) *Enchantments of Modernity: Empire, Nation, Globalization* (London: Routledge), pp. 291–309.

—— (2015) "Totalitarian tears: does the crowd really mean it?" *Cultural Anthropology* 30(1): 91–112.

McDowell, L. (2009) *Working Bodies: Interactive Service Employment and Workplace Identities* (Hoboken: Wiley-Blackwell).

McLean, S. (2007) "Introduction: why imagination?" *Irish Journal of Anthropology* (Special issue, *Engaging Imagination: Anthropological Explorations in Creativity*) 10(2): 5–10.

McMichael, W.H. (2002) "Sex slaves," *Navy Times*, August 12. Available at: http://www.vvawai.org/archive/general/sex-slaves.html (accessed July 9, 2015).

Meyer, B. (2011) "Mediation and immediacy: sensational forms, semiotic ideologies and the question of the medium," *Social Anthropology* 19: 23–39.

Min, S.J. (2002) "Anti-U.S. focus marks weekend rallies," *JoongAng Daily*, December 16. Available at: http://joongangdaily.joins.com/article/view.asp?aid=1912435 (accessed July 9, 2015).

Mitchell, W.J.T. (2005) *What do Pictures Want? The Lives and Loves of Images* (Chicago: University of Chicago Press).

Miyake, M. (1996) "Japan's encounter with Germany, 1860–1914: an assessment of the German legacy in Japan," *European Legacy* 1: 245–249.

Miyazaki, H. (2004) *The Method of Hope: Anthropology, Philosophy, and Fijian Knowledge* (Stanford, CA: Stanford University Press).

—— (2006) "Economy of dreams: hope in global capitalism and its critiques," *Cultural Anthropology* 21(2): 147–172.

Moon, C.I. and Lee, S. (2010) "Military spending and the arms race on the Korean peninsula," *Asia-Pacific Journal: Japan Focus*, March 28. Available at: http://www.japanfocus.org/-Chung_in-Moon/3333 (accessed July 9, 2015).

Moon, K.H.S. (1997) *Sex Among Allies: Military Prostitution in U.S.–Korea Relations* (New York: Columbia University Press).

—— (1999) "South Korean movements against militarized sexual labor," *Asian Survey* 34(2): 310–327.

Moon, S. (2005) *Militarized Modernity and Gendered Citizenship in South Korea* (Durham, NC: Duke University Press).

—— (2010a) "Camptown prostitution and the imperial SOFA: abuse and violence against transnational camptown women in South Korea," in Hoehn, M. and Moon, S. (eds.) *Over There: Living with the U.S. Military Empire from World War Two to the Present* (Durham, NC: Duke University Press), pp. 337–365.

—— (2010b) "In the U.S. Army but not quite of it: contesting the imperial power in a discourse of KATUSAS," in Hoehn, M. and Moon, S. (eds.) *Over There: Living with the U.S. Military Empire from World War Two to the Present* (Durham, NC: Duke University Press), pp. 231–257.

—— (2010c) "Regulating desire, managing the empire: US military prostitution in South Korea, 1945–1970," in Hoehn, M. and Moon, S. (eds.) *Over There: Living with the U.S. Military Empire from World War Two to the Present* (Durham, NC: Duke University Press), pp. 39–77.

Mosse, G.L. (1988) *Nationalism and Sexuality: Middle-class Morality and Sexual Norms in Modern Europe* (Madison: University of Wisconsin Press).

Murillo, D. and Sung, Y. (2013) "Understanding Korean capitalism: chaebols and their corporate governance," ESADEgeo, Position Paper No. 33. Barcelona: ESADE.

Myers, B.R. (2010) *The Cleanest Race: How North Koreans See Themselves – And Why it Matters* (New York: Melville House Publishing).

Narotzky, S. and Besnier, N. (2014) "Crisis, value, and hope: rethinking the economy. An introduction to supplement 9," *Current Anthropology* 55(S9): S4–S16.

Neff, R. (2010) "Kenneth L. Markle: sadistic murderer or scapegoat?" *The Marmot's Hole* (online blog) February 24. Available at: http://www.rjkoehler.com/2010/02/24/kenneth-markle-was-he-innocent/ (accessed July 9, 2015).

Network of Concerned Anthropologists (2009) *The Counter-counterinsurgency Manual: Or, Notes on Demilitarizing American Society* (Chicago: Prickly Paradigm Press).

Onishi, N. (2003) "Korean actor's reality drama: coming out as gay," *The New York Times* October 1. Available at: http://www.nytimes.com/2003/10/01/world/korean-actor-s-reality-drama-coming-out-as-gay.html (accessed July 9, 2015).

Ortner, S. (1978) "The virgin and the state," *Feminist Studies* 4(3): 19–35.

—— (1984) "Theory in anthropology since the sixties," *Comparative Studies in Society and History* 26(1): 126–166.

Pae, K.C. (2014) "Feminist activism as interfaith dialogue: a lesson from Gangjeong village of Jeju Island, Korea," *Journal of Korean Religions* 5(1): 55–69.

Paik, N. (2000) "Coloniality in Korea and a South Korean project for overcoming modernity," *Interventions. International Journal of Postcolonial Studies* 2(1): 73–86.

—— (2005) "How to assess the Park Chung Hee Era and Korean Development," *The Asia-Pacific Journal: Japan Focus*, December 29. Available at: http://old.japanfocus.org/-Paik_Nak_chung/1725 (accessed July 9, 2015).

—— (2009) "Korea's division system and its regional implications" (lecture, Australian National University, August 25, 2009). Available at: http://en.changbi.

com/2009/09/03/paik-nak-chung-koreas-division-system-and-its-regional-implications/ (accessed July 9, 2015).

Park, B. (2014) "Delay of OPCON transfer could reverse relocation of US troops from Seoul," *Hankyoreh*, September 19. Available at: http://english.hani.co.kr/arti/english_edition/e_international/655934.html (accessed July 9, 2015).

Park, J. (2014) "Former Korean 'comfort women' for U.S. troops sue own government," *Reuters*, July 11. Available at: http://uk.reuters.com/article/2014/07/11/uk-southkorea-usa-military-idUKKBN0FG0WK20140711 (accessed July 9, 2015).

Park, M. (2005) "Organizing dissent against authoritarianism: the South Korean student movement in the 1980s," *Korea Journal* 45(3): 261–289.

Park, N.J. (2008) "Redesigning Korea," *Hankyoreh*, June 19. Available at: http://english.hani.co.kr/arti/english_edition/e_opinion/294163.html (accessed July 9, 2015).

Park, W. (2002) "The unwilling hosts: state, society, and the control of guest workers in South Korea," in Yaw, D. (ed.) *Migrant Workers in Pacific Asia* (London: Frank Cass), pp. 67–95.

Park-Kim, S., Lee-Kim, S., and Kwon-Lee, E. (2007) "The lesbian rights movement and feminism in South Korea," *Journal of Lesbian Studies* 10(3–4): 161–190.

Patten, E. and Parker, K. (2011) "Women in the U.S. military: growing share, distinctive profile," *PEW Social and Demographic Trends*, December 22. Available at: http://www.pewsocialtrends.org/files/2011/12/women-in-the-military.pdf (accessed July 9, 2015).

Pine, F. (2014) "Migration as hope: space, time, and imagining the future," *Current Anthropology* 55(S9): S95–S104.

Pinney, C. (2011) *Photography and Anthropology* (London: Reaktion).

Powers, R. (n.d.) "Installation overview – United States Army Garrison (USAG) Yongsan, Korea." Available at: http://usmilitary.about.com/od/armybaseprofiles/ss/Yongsan.htm#showall (accessed July 12, 2015).

Price, D. (2008) *Anthropological Intelligence* (Durham, NC: Duke University Press).

—— (2011) *Weaponizing Anthropology* (Oakland, CA: CounterPunch and AKPress).

Pyke, K.D. (1996) "Class-based masculinities: the interdependence of gender, class, and interpersonal power," *Gender & Society* 10(5): 527–549.

Pyle, K.B. (1996) *The Making of Modern Japan* (Lexington, MA: D.C. Heath).

Rabiroff, J. (2009) "Philippines takes aim at juicy bar trafficking," *Stars and Stripes* , November 15. Available at: http://www.stripes.com/news/philippines-takes-aim-at-juicy-bar-trafficking-1.96490 (accessed July 9, 2015).

Ramstad, E. (2011) "Rumsfeld seized Roh's election to change alliance," *Korea Real Time* (*Wall Street Journal* blog), February 8. Available at: http://blogs.wsj.com/korearealtime/2011/02/08/rumsfeld-seized-rohs-election-to-(accessed July 9, 2015).

Rauhala, E. (2010) "South Korea: should foreign teachers be tested for HIV?" *Time Magazine*, December 24. Available at: http://www.time.com/time/world/article/0,8599,2039281,00.html#ixzz19xJi5ub0 (accessed July 9, 2015).

Renan, E. (2001) "What is a nation?" in Pecora, V. (ed.) *Nations and Identities: Classic Readings* (Malden, MA: Blackwell), pp.162–176.

Rinser, L. and Yun, I. (1977) *Der verwundete Drache. Dialog über Leben und Werk des Komponisten* (Frankfurt/Main: S. Fischer).

Roehner, B.M. (2007) *Relations between Allied Forces and The Population of Japan, 15 August 1945–31 December 1960* (Paris: University of Paris).

—— (2014) *Relations between U.S. Forces and the Population of South Korea, 1945–2010* (Institute for Theoretical and High Energy Physics, University of Paris 6, Working Report). Available at: http://www.lpthe.jussieu.fr/~roehner/ock.pdf (accessed July 9, 2015).

Rowland, A. (2010) "South Korea bar district offers a safe haven for gay service members," *Stars and Stripes*, March 14. Available at: http://www.stripes.com/news/south-korea-bar-district-offers-a-safe-haven-for-gay-servicemembers-1.99964 (accessed July 9, 2015).

—— (2014) "Cavalry battalion arrives in Korea for 'plus-up' deployment," *Stars and Stripes*, January 30. Available at: http://www.stripes.com/news/pacific/cavalry-battalion-arrives-in-korea-for-plus-up-deployment-1.264671#.UzjZnly9Y8M (accessed July 9, 2015).

Russell, M. (2008) *Pop Goes Korea: Behind the Revolution in Movies, Music, and Internet Culture* (Berkeley: Stone Bridge Press).

Ryan, D. (2012) "Anti-Americanism in Korean films," *Colorado Journal of Asian Studies* 1(1): 94–109.

Sahlins, M. (2005) "Structural work: how microhistories become macrohistories and vice versa," *Anthropological Theory* 5(1): 5–30.

Salazar, N.B. (2011) "The power of imagination in transnational mobilities," *Identities: Global Studies in Culture and Power* 18(6): 576–598.

Salazar, N.B. (2012) "Tourism imaginaries: a conceptual approach," *Annals of Tourism Research* 39(2): 863–882.

Santos, A.F. (1992) "Gathering the dust: the bases issue in the Philippines," in Sturdevant, S.P. and Stoltzfus, B. (eds.) *Let the Good Times Roll: Prostitution and the U.S. Military in Asia* (New York: The New Press), pp. 32–44.

Sartre, J.P. (2004 [1940]) *The Imaginary: A Phenomenological Psychology of the Imagination* (London: Routledge).

Savada, A.M. and Shaw, W. (eds.) (1990) *South Korea: A Country Study* (Washington: GPO for the Library of Congress). Available from http://countrystudies.us/south-korea/ (accessed July 9, 2015).

Schmid, A. (2002) *Korea between Empires, 1895–1919* (New York: Columbia University Press).

Schober, E. (2007) "Trafficking – what is that?" *Focaal* 49: 124–128.

Schober, E. (2010) "Subverting the military normal," *Focaal* 58: 109–114.

Schuessler, R. (2015) "Korean-American military brides find refuge in tiny Missouri church," *Aljazeera America*, March 23. Available at: http://america.aljazeera.com/

articles/2015/3/21/missouri-community-korean-american-military-brides.html (accessed July 11, 2015).

Scofield, D. (2004) "The mortician's tale: time for the US to leave Korea," *Asia Times*, January 28. Available at: http://www.atimes.com/atimes/Korea/FA28Dg02.html (accessed July 9, 2015).

Seo, D. (2001) "Mapping the vicissitudes of homosexual identities in South Korea," *Journal of Homosexuality* 40(3–4): 65–78.

Severi, C. (2015) *The Chimera Principle: An Anthropology of Memory and Imagination* (Chicago: University of Chicago Press).

Shaw, M. (2012) "Twenty-first-century militarism: a historical-sociological framework," in Stavrianakis, A. and Selby, J. (eds.) *Militarism and International Relations: Political Economy, Security, Theory* (London: Routledge).

Shefer, T. and Mankayi, N. (2007) "The (hetero)sexualization of the military and the militarization of (hetero)sex: discourses on male (hetero)sexual practices among a group of young men in the South African military," *Sexualities* 10(2): 189–207.

Shigematsu, S. and Camacho, K.L. (2010) *Militarized Currents: Toward a Decolonized Future in Asia and the Pacific* (Minneapolis: University of Minnesota Press).

Shim, D. (2009) *A Shrimp amongst Whales? Assessing South Korea's Regional-Power Status*, German Institute for Global and Area Studies, Working Paper. Available at http://www.giga-hamburg.de/dl/download.php?d=/content/publikationen/pdf/wp107_shim.pdf (accessed July 9, 2015).

Shin, G.W. (2006) *Ethnic Nationalism in Korea: Genealogy, Politics, and Legacy* (Stanford, CA: Stanford University Press).

Shin, J. (2005) "Invasion of privacy degrades Korean women twice over," *Chosun Ilbo*, January 24. Available at: http://english.chosun.com/site/data/html_dir/2005/01/24/2005012461027.html (accessed July 9, 2015).

Shin, J., Ostermann, C.F., and Person, J. (2013) *North Korean Perspectives on the Overthrow of Syngman Rhee, 1960*, NKIDP E-Dossier No. 13, Woodrow Wilson International Center for Scholars. Available at: http://www.wilsoncenter.org/sites/default/files/NKIDP_eDossier_13_North_Korean_Perspectives_on_the_Overthrow_of_Syngman_Rhee.pdf (accessed July 9, 2015).

Shin, K.Y. (2006) "The citizens' movement in Korea," *Korea Journal* 46(2): 5–34.

Shorrock, T. (1999) "Kwangju Diary: the view from Washington," in Lee, J. (ed.) *Kwangju Diary: Beyond Death, Beyond the Darkness of the Age* (Los Angeles: The UCLA Asian Pacific Monograph Series), pp. 151–172.

Simbulan, R.G. (2009) "People's movement responses to evolving U.S. military activities in the Philippines," in Lutz, C. (ed.) *The Bases of Empire: The Global Struggle against U.S. Military Posts* (London: Pluto), pp. 145–180.

Sjoberg, L. and Via, L. (eds.) (2010) *Gender, War, and Militarism* (Santa Barbara, CA: Greenwood Publishing Group).

Skinner, J. and Theodossopoulos, D. (2011) *Great Expectations: Imagination and Anticipation in Tourism* (Oxford: Berghahn Books).

Slavin, E. (2011) "South Korea to review status of forces agreement with U.S," *Stars and Stripes*, October 12. Available at: http://www.stripes.com/promotions/2.1066/pacific/korea/s-korea-to-review-status-of-forces-agreement-with-u-s-1.157548 (accessed July 9, 2015).

Slavin, E. and Hwang H. (2007) "Arrest warrant issued for GI in rape case," *Stars and Stripes*, January 19. Available at: http://stripes.com/article.aspsection=104&article=41853&archive=true (accessed July 9, 2015).

Smith, N. (1982) "Gentrification and uneven development," *Economic Geography* 58(2): 139–155.

—— (1986) *Gentrification of the City* (Boston: Unwin Hyman).

—— (2002) "New globalism, new urbanism: gentrification as global urban strategy," *Antipode* 34: 427–450.

Soh, C.S. (1996) "The Korean 'comfort women': movement for redress," *Asian Survey* 36(12): 1226–1240.

—— (2009) *The Comfort Women: Sexual Violence and Postcolonial Memory in Korea and Japan* (Chicago: University of Chicago Press).

Song, W. (2010) 'Crying Nut on getting older and getting naked," *Korea Herald*, March 30. Available at: *http://www.koreaherald.com/national/Detail.jsp?newsMLId=20090828000013* (accessed July 9, 2015).

Stars and Stripes (2009) 'Itaewon: what's in a name?," *Stars and Stripes*, January 4. Available at: http://www.stripes.com/news/itaewon-what-s-in-a-name-1.86704 (accessed July 9, 2015).

Stavrianakis, A. and Selby, J. (2012) 'Militarism and international relations in the 21st Century," in Stavrianakis, A. and Selby, J. (eds.) *Militarism and International Relations: Political Economy, Security, Theory* (London: Routledge), pp. 3–28.

Stiehm, J. (1996) *It's Our Military, Too! Women and the U.S. Military* (Philadelphia, PA: Temple University Press).

Strathern, A., Stewart, P.J., and Whitehead, N.L. (2006) *Terror and Violence: Imagination of the Unimaginable* (London: Pluto Press).

Sturdevant, S.P. and Stoltzfus, B. (eds) (1992) *Let the Good Times Roll: Prostitution and the U.S. Military in Asia* (New York: The New Press).

Talusan, M. (2015) 'How the killing of a trans Filipina woman ignited an international incident," *Vice Magazine*, February 23. Available at: http://www.vice.com/read/the-trial-of-a-us-marine-accused-of-killing-a-trans-filipina-woman-starts-today-will-he-go-free (accessed July 9, 2015).

Tanaka, Y. (2001) *Japan's Comfort Women* (London: Routledge).

Taylor, C. (2003) *Modern Social Imaginaries* (Durham, NC: Duke University Press).

Tikhonov, V. (2003) "World is a battlefield: Social Darwinism as the new world model of Korean intelligentsia of the 1900s," *Bochumer Jahrbuch zur Ostasienforschung* 27: 86–106.

—— (2009) "Militarism and anti-militarism in South Korea: "militarized masculinity" and the conscientious objector movement," *Asia-Pacific Journal:*

Japan Focus, March 16. Available at: http://www.japanfocus.org/-Vladimir-Tikhonov/3087/article.html (accessed July 9, 2015).

—— (2010) *Social Darwinism and Nationalism in Korea: The Beginnings, 1880s–1910s – Survival as an Ideology of Korean Modernity* (Leiden: Brill).

Tong, K.W. (1991) "Korea's forgotten atomic bomb victims," *Bulletin of Concerned Asian Scholars* 23: 31–37.

Tsing, A. (2000) "The global situation," *Cultural Anthropology* 55(3): 327–360.

—— (2005) *Friction: An Ethnography of Global Connection* (Princeton, NJ: Princeton University Press).

Turnbull, J. (2009) "Downturn spawns 'flower men' wind," *Korea Times*, April 3. Available at: http://www.koreatimes.co.kr/www/news/nation/2009/04/117_42550.html (accessed July 9, 2015).

Turner, V. (1967) *The Forest of Symbols: Aspects of Ndembu Ritual* (Ithaca, NY: Cornell University Press).

—— (1969) *The Ritual Process: Structure and Anti-structure* (New York: Aldine de Gruyter).

Vagts, A. (1937) *A History of Militarism* (New York: W.W. Norton).

Visweswaran, K. (ed.) (2014) *Everyday Occupations: Experiencing Militarism in South Asia and the Middle East* (Philadelphia, PA: University of Pennsylvania Press).

Wagner, B. and VanVolkenburg, M. (2012) "HIV/AIDS tests as a proxy for racial discrimination? A preliminary investigation of South Korea's policy of mandatory in-country HIV/AIDS tests for its foreign English teachers," *Journal of Korean Law* 11: 179–245.

Walhain, L. (2007) "Transcending Minjok: how redefining nation paved the way to Korean democratization," *Studies on Asia* 4(2): 84–101.

Weber, M. (1991) *From Max Weber: Essays in Sociology* (New York: Routledge).

Weitzer, R. (ed.) (2000) *Sex for Sale: Prostitution, Pornography, and the Sex Industry* (New York: Routledge).

—— (2005) "Flawed theory and method in studies of prostitution," *Violence Against Women* 11(7): 934–949.

Whitehead, N.L. (2004) *Violence* (Oxford: James Currey / SAR Press).

Wickham, J.A. (2000) *Korea on the Brink: A Memoir of Political Intrigue and Military Crisis* (Dulles: Potomac Books).

Williamson, G. (2004) "Review – modern social imaginaries," *Metapsychology Online Reviews* 8(45). Available at: http://metapsychology.mentalhelp.net/poc/view_doc.php?type=book&id=2376 (accessed 9 July 2015).

Winter, B. (2011) "Guns, money and justice: the 2005 Subic rape case," *International Feminist Journal of Politics* 13(3): 371–389.

Witworth, S. (2004) *Men, Militarism, and UN Peacekeeping* (Boulder, CO: Lynne Rienner).

Yang, H. (1998) "Re-membering the Korean military comfort women: nationalism, sexuality, and silencing," in Kim, E.H. and Choi, C. (eds.) *Dangerous Women: Gender and Korean Nationalism* (London: Routledge), pp. 123–140.

Yea, S. (2002) "Rewriting rebellion and mapping memory in South Korea: the (re) presentation of the 1980 Kwangju uprising through Mangwol-dong cemetery," *Urban Studies* 39(9): 1551–1572.

—— (2005) "Labour of love – Filipina entertainer's narratives of romance and relationships with GIs in U.S. military camp towns in Korea," *Women's Studies International Forum* 28: 456–427.

Yeo, A. (2006) "Local–national dynamics and framing in South Korean anti-base movements," *Kasarinlan: Philippine Journal of Third World Studies* 21(2): 34–60.

—— (2010) "Anti-base movements in South Korea: comparative perspective on the Asia-Pacific," *Asia-Pacific Journal: Japan Focus*, June 14. Available at: http:// japanfocus.org/-Andrew-Yeo/3373/article.html (accessed July 9, 2015).

Yoshimi, Y. (2002) *Comfort Women: Sexual Slavery in the Japanese Military during World War II, Asia Perspectives* (New York: Columbia University Press).

Yuh, J. (2002) *Beyond the Shadow of Camptown: Korean Military Brides in America* (New York: New York University Press).

Yuval-Davis, N. (1997) *Gender and Nation* (London: Sage).

Index

Italicized page numbers refer to illustrations. An "n" after a page number indicates a note, with the note number following.